ISSUES IN OCEANOGRAPHY

Daniel C. Abel
Coastal Carolina University

Robert L. McConnell
Mary Washington College

Eric Koepfler
Coastal Carolina University

Prentice Hall
Upper Saddle River, New Jersey 07458

Library of Congress Cataloging-in-Publication Data

Abel, Daniel C.
  Issues in oceanography / Daniel C. Abel, Robert L. McConnell, Eric Koepfler.
    p.   cm.
  Includes bibliographical references and index.
  ISBN 0-13-018603-1 (pbk.)
  1. Oceanography. I. McConnell, Robert L. II. Koepfler, Eric. III. Title.

GC11.2 .A33 2001
551.46—dc21   00–048362

Senior Editor: *Patrick Lynch*
Production Editor: *Shari Toron*
Art Director: *Jayne Conte*
Manufacturing Manager: *Trudy Pisciotti*
Manufacturing Buyer: *Michael Bell*
Photo Editor: *Beth Boyd*
Marketing Manager: *Christine Henry*
Assistant Editor: *Amanda Griffith*
Editorial Assistant: *Sean Hale*
Cover Designer: *Bruce Kenselaar*

 © 2001 by Prentice-Hall, Inc
Upper Saddle River, New Jersey 07458

Printed in the United States of America.

10 9 8 7 6 5 4 3 2 1

ISBN 0-13-018603-1

Prentice-Hall International (UK) Limited, *London*
Prentice-Hall of Australia Pty. Limited, *Sydney*
Prentice-Hall Canada, Inc., *Toronto*
Prentice-Hall Hispanoamericana, S.A., *Mexico*
Prentice-Hall of India Private Limited, *New Delhi*
Prentice-Hall of Japan, Inc., *Tokyo*
Pearson Education Asia Pte. Ltd.
Editora Prentice-Hall do Brasil, Ltda., *Rio de Janeiro*

ISSUES IN
OCEANOGRAPHY

Charles Seale-Hayne Library
University of Plymouth
(01752) 588 588
LibraryandITenquiries@plymouth.ac.uk

Contents

About the Authors

Daniel C. Abel

Daniel C. Abel, is a native of Charleston, South Carolina. He earned his Ph.D.in 1986 in Marine Biology from Scripps Institution of Oceanography, University of California, San Diego, and was a post doctoral fellow in the Marine Biomedical Research Program of the Medical University of South Carolina. He is also co-author of *Environmental Issues: Measuring, Analyzing, and Evaluating*. He currently teaches in the Marine Science Department at Coastal Carolina University, where his research focuses on the sharks of Winyah Bay. A devotee of the comic strip *This Modern World*, he resides in Pawley's Island, South Carolina, with his wife Mary, daughter Juliana, and son Louis, and is active in local environmental issues.

Robert L. McConnell

Robert L. McConnell is Professor of Geology and Environmental Science at Mary Washington College in Fredericksburg, Virginia, and lives in Arlington, across the Potomac from Washington D.C. He spent a year as Lecturer at the Open University in England, and still tries to spend time each year studying aspects of U.K. geology in the magnificent exposures along the country's west coast. He has also studied geology in Australia and Iceland. He has published articles on environmental geology and energy policy in scholarly journals, given numerous talks at scientific meetings and conferences, and has written more than two dozen commentary articles for major newspapers on the same topics. He is also co-author of *Environmental Issues: Measuring, Analyzing and Evaluating*, published in 1999 by Prentice Hall.

Eric Koepfler

Eric Koepfler obtained his Ph.D. from the Virginia Institute of Marine Science, College of William & Mary in 1989. He completed a postdoctoral fellowship at the University of Texas, Marine Science Institute in 1990. He is presently an Associate Professor in the Marine Science Department at Coastal Carolina University. Dr. Koepfler has conducted research through the National Science Foundation (NSF), the National Oceanographic and Atmospheric Administration (NOAA), and the National Aeronautics and Space Administration (NASA). Dr. Koepfler has also been an officer in the Southeastern Estuarine Research Society, a member organization of the Estuarine Research Federation (ERF) which studies processes that occur in estuarine and coastal environments. Eric lives in Myrtle Beach, South Carolina with his wife Julia and children Allyson and Seth.

Preface

To the Professor

This text contains twelve "Issues in Oceanography," brief projects which use pressing marine environmental issues as a means to develop your students' critical thinking skills in a deliberate and structured way. By their nature, they require students to integrate topics from across sub-disciplines to measure, analyze, and evaluate the issue using the discipline and method of a scientist. The text consists of brief introductions. The remainder of the analysis is found on the Companion Web Site at www.prenhall.com/oceanissues.

These Issues, like our previous text, *Environmental Issues: Measuring, Analyzing, and Evaluating*, grew out of our desire to encourage in the reader a more "active" style of learning and discourage what we see as a passive and unhealthy dependency on the faculty person as the "expert." One of our major objectives is to help develop math literacy (numeracy) among today's students; not necessarily arcane math, but the kind of math needed to properly quantify environmental issues. Such skills involve the ability to manipulate large numbers using scientific notation and exponents, the ability to use compound growth equations containing natural logs, etc. This lack of math skills often leaves students unprepared to deal with the complexity of today's environmental issues. For a more detailed version of our methods and a different set of issues, we refer you to *Environmental Issues*. In this modest text, we provide detailed introductions for each of our twelve topics. The Issues projects are designed to take from one to three hours to complete. The Instructor's Manual contains suggestions for employing the Issues in your course, as well as test questions based on their content.

The Issues and critical thinking questions have been designed to be provocative. Although we have made the content as factual as possible, we admit to having strong convictions about these issues. Convictions are not, however, biases. Our views as scientists are subject to change as evidence supporting our convictions changes. Indeed this aspect can be turned to a major advantage. Ask your students to look for examples of bias in the questions, and then discuss with them the difference in science between "bias" and "conviction." No doubt it will prove a fruitful activity, and may lead students into research (perhaps to "prove us wrong"), which is the essence of progress in the search for scientific truth.

To the Student

As environmental scientists, we care deeply about the health of the ocean, and we feel that you, as a responsible citizen who will have to make increasingly difficult choices in the years ahead, need to be concerned about them as well. We hope you will find the Issues in Oceanography contained in this text to be a provocative introduction to a number of these issues, including many that you may have never even thought about. Contained in the text are brief introductions. The complete projects are found on the Companion Web Site at www.prenhall.com/oceanissues. These are real-life issues, not hypothetical ones, and you need certain basic skills to fully understand them.

They are as follows:

- You must be familiar with and be able to use the units of the metric system.
- You must be able to use a few simple mathematical formulas to quantify the issues you will be debating, and you must be able to carry out the calculations accurately.
- You must rigorously and continuously assess your thinking and apply certain critical thinking skills and techniques when discussing the implications of your calculations.

We understand that many students have some "math anxiety," so we use a step-by-step method to take you through many of the calculations in the Issues. Math proficiency is one of the important skills necessary for fully understanding environmental issues, and without these skills, your only option is to make choices on the basis of which "expert" you believe. But becoming educated is much more than simply acquiring skills. Therefore, we have another two fundamental objectives: to provide you with the knowledge and intellectual standards necessary to apply critical thinking to environmental studies, and to foster your ability to critically evaluate issues.

How to Use this Book

The content of Issues in Oceanography resides only in part between the covers of the text. The remainder is contained on the web site www.prenhall.com/oceanissues. For each of the issues, you should begin by reading the 2-4 page introductions in the book. Then access our web site and begin your own analysis of the issue.

Each web site begins with an "Issue Discussion." An "Analysis" section consisting of relevant background information, questions, and spaces for your answers follows this. You can submit your responses to your instructor by clicking the appropriate button. In many of the issues we have provided you with a "Media Analysis" section where you can listen to short clips of interviews or news reports and answer questions about them. Finally, each site contains an annotated list of links, which we call "Destinations."

What is Critical Thinking?

Critical thinking involves using a set of criteria and standards by which the reasoner constantly assesses her/his thinking. At the core of critical thinking is self-assessment, and here are the standards we all need to use in order to reason effectively.

First, all reasoning is an attempt to figure something out, or to solve a problem.

Second, the critical thinker above all must be clear on the nature of the problem to be solved, which introduces what some have called the "gateway" critical thinking standard: clarity. For if a statement is not clear, it cannot be studied or assessed effectively. Think about your own career as a student for a moment: have there been times when you did not clearly understand an assignment or a text question, which in turn contributed to your missing the question? Have there been times when, in discussing an assignment with some of your classmates, that several versions of the assignment emerged? If so, you have an illustration of the importance of clarity in thinking.

Next (but in no particular order) the reasoner uses evidence, consisting of the results of experiments, that is data or information, and observations to solve a problem. This introduces other essential standards. We must be sure that the evidence is relevant to the issue being analyzed, and that the information is sufficiently accurate (true) and precise (sufficiently detailed) to use to solve our problem. Do you see now why critical thinkers must constantly assess their thinking? This raises another issue: critical thinking is hard work--it isn't easy!

Other criteria we must use to assess our reasoning are: we must ensure that our thinking is sufficiently broad and deep, such that we have considered all reasonable information; we must ensure that we have been fair-minded— that we have fairly considered all reasoned points of view if we are trying to solve a problem that involves reasoned judgment. An example: How much pollution can we put into the ocean without damaging marine life?

In summary we must constantly assess our thinking when we are trying to solve a problem, when we are using reasoned judgment. We must be sure that we clearly understand the problem, we must use relevant information in the solution, the information must be sufficiently broad and deep, and we must consider fairly all relevant points of view, that is, points of view that obtain from reasoning--not just "opinions." Finally here, note that there are significant differences between "opinions" and "points of view" in our lexicon. Anyone can have an opinion, but for it to be relevant to scientific inquiry it must be an opinion informed by the principles of critical thinking and logic. For more information on this topic, go to our earlier book Environmental Issues, or check out the Center for Critical Thinking's web site at www.criticalthinking.org

Conclusion

As our national and global population grows and changes and our relationships with other nations and peoples evolve, environmental issues, especially those involving our common heritage, the oceans and the atmosphere, will become increasingly complicated. We hope that you will be challenged by the issues discussed in this text and that you will research them and become an "expert" on the topics yourself. In fact, if we may be allowed a hidden agenda, this is it.

Reviewers

The authors would like to extend a special thanks to colleagues who reviewed portions of this text and offered many helpful comments and suggestions.

William M. Landing
  Florida State University
Lawrence Krissek
  Ohio State University
William Ellis
  Maine Maritime Academy
Kathleen M. Browne
  Rider University
A. Quinton White
  Jacksonville University
Rodney M. Feldmann
  Kent State University
George P. Burbanck
  Hampton University
Jane Matty
  Central Michigan University
Lenore Tedesco
  Indiana University–Purdue University Indianapolis

Coastal Population Growth: A Global Ecosystem at Risk

- How does human population growth threaten coastal areas?
- How can we measure these threats?
- What can we do about them?
- Who is responsible for solving the problem?

Introduction

Chances are, you live within 100 km (62 mi) of the beach. According to the World Resources Institute[1], at least 60 % of the planet's human population lives that close to the coastline. Coastal areas have the fastest growing populations as well. Not surprisingly, over half the world's coastlines are at significant risk from development-related activities. Some of these activities are:

- Conversion of tropical mangrove communities to fish and shrimp farms
- Expansion of many coastal cities which are in the direct path of hurricanes, monsoons, and tropical storms
- Pollution from Western-style industrial agriculture, including industrial meat production
- Releases of untreated or partially treated human sewage (between one-third and two-thirds of the human waste generated in developing countries is not even collected).

Coastal cities are reaching sizes unprecedented in human history. Here are but a few examples: Sao Paolo, Brazil, 17 million; Bombay, India, 16.5 million; Lagos, Nigeria, 11 million (and growing at a rate of 5.7% per year); Dhaka, Bangladesh, 8 million (growth rate nearly 6% per year). Most of these cities make little attempt to treat the sewage produced by their residents. On our web site, we evaluate the impact of the growth of some of these cities.

Sewage and Animal Waste Impacts

Sewage and animal waste may contain a number of harmful substances, including pathogens such as bacteria and viruses.

In addition to pathogens, untreated or partially treated sewage contains high concentrations of nitrogen and phosphorus compounds, which are essential nutrients for growth of plants and algae. Excess nitrogen and phosphorus can stimulate rapid growth of algae, which can smother organisms attached to the bottom and can also form a thick layer at the water surface. This surface layer of algae may thrive in the short term, but it can block sunlight from lower layers, whose plants eventually die. More-

over, rapidly-growing algae can quickly deplete nutrients, resulting in rapid die-offs. In both cases, oxygen-using bacteria begin to decompose the dead algae. This can cause hypoxia (dangerously low oxygen levels) or anoxia (absence of oxygen).

Sewage can also contain heavy metals such as copper and zinc, which are toxic to plants and algae (these metals are often fed to pigs as supplements in amounts that exceed the animals' ability to metabolize them, so they end up in the animals' waste).

Case Study: Bangladesh

Bangladesh (Figure 1), a country about the size of Illinois, Wisconsin, or Florida, lies on the northern shore of the Indian Ocean. Dr. Selina Begum of the U.K.'s Bradford University wrote[2]:

> "Bangladesh is a delta of the Ganges, Brahmaputra and Meghna Rivers. Tributaries and distributaries [branches] of the river system cover all of the country. The rivers rise in the Himalayas and drain a catchment area of about 1.5 million km^2 [577,000 mi^2], only 7.5 percent of which lies in Bangladesh [our emphasis. Stop and explain why the responsibility for Bangladesh's destiny may lie with others far away.]."
>
> "The country is prone to meteorological and geologic natural disasters, due to its geographic location, climate, variable topography, dynamic river system and exposure to the sea. The steady increase in population continuously increases the potential for natural disaster."

Bangladesh thus makes an ideal case illustrating the impacts of coastal growth.

Background

Bangladesh was originally part of Pakistan after the Indian subcontinent gained independence from Britain in 1947. In 1970, Bangladesh seceded from the rest of Pakistan. It had a 1999 population of 127 million on a land area of 144,000 km^2 (55,000 mi^2).

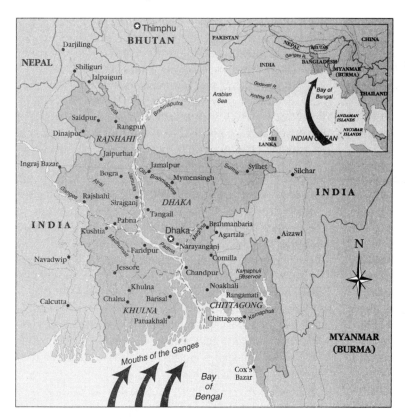

Figure 1

Map of Bangladesh Showing Generalized Tracks of Twentieth
Century Cyclones.

Figure 2

Bangladesh countryside following the 1991 cyclone
that killed approximately 145,000 people. *(Liason
Agency, Inc.)*

Bangladesh is vulnerable to catastrophic flooding from
river discharges as well as from tropical storms. In addition, it is defenseless against sea level rise from global climate change since more than half the country lies at an el-
evation of less than 8 meters (26 ft) above sea level. As a
result, about 25–30% of the country is flooded each year,
which can increase to 60–80% during major floods. Moreover, 17 million people live on land that is less than 1 m
(3.3 ft) above sea level.

Flooding in Bangladesh can result from some combination of the following: (1) excess rainfall and snowmelt
in the catchment basin, especially in the foothills of the
Himalayas (an area where many of the Earth's rainfall
records were set); (2) simultaneous peak flooding in all
three main rivers; and (3) high tides in the Bay of Bengal, which can dam runoff from rivers and cause the water
to "pond up" and overtop its banks.

In addition, changes in land use in the far reaches of the
catchment area can result in profound changes in the flood
potential of Bangladesh's rivers. For example, deforestation of the Himalayan foothills for agriculture, fuelwood,
or human habitation can make the hills more prone to flash
flooding. Then, during the torrential monsoon rains, mud-
slides can dump huge volumes of sediment into the rivers.
This sediment can literally pile up in the river channels,
reducing channel depth and thereby the channel's capac-

ity to transport water. Result: more water sloshing over the
rivers' banks during floods, and more disastrous floods
in Bangladesh. Another result: the silt can smother offshore reefs when it finally reaches the ocean. Geologists
estimate that the drainage system dumps at least 635 million metric tons (1.4×10^{12} lb) of sediment a year into the
Indian Ocean, which is four times the amount of the present Mississippi River.

Some researchers feel that Western-style development is
contributing to Bangladesh's increased vulnerability to natural disasters. Farhad Mazhar, director of the Bangladeshi
research and development group Ubinin, concludes that the
basic problem *is* Western-style development. He cites one
example: During pre-development days Bangladeshi vil-
lagers depended on boats for transportation during the monsoon season. But, "to be modern" the Bangladeshis built
roads, which blocked floodwaters. These and other "modern improvements" increased flooding from cyclones and
have also increased the damage and loss of life. Mazhar concludes,"… if you compare the history of the old cyclones
and flood waters, you'll see… the misery has increased with
this kind of 'engineering' solution to water[3]."

Truly catastrophic floods can result when high tides co-
incide with tropical storms (Figure 2). In the early 1990s,
such a combination took over 125,000 lives. About one tenth
of all Earth's tropical cyclones occur in the Bay of Bengal

(Figure 1), and about 40% of the global deaths from storm surges occur in Bangladesh.

Storm surges[4] result when cyclones with winds that can exceed 250 km/hour (155 mi/hr), move onshore, and push a massive wall of seawater onshore with them. Surges in excess of 5 meters (16.4 ft) above high tide are not uncommon, so you can imagine the impact such a storm can have on a country in which half the land is within 8 meters (26.2 ft) of sea level!

Computer models forecast an average rise of sea level of 66 cm (26 in) by 2100 due to global warming, but sea level could rise anywhere from 20–140 cm (8–55 in). There is, however, a complicating factor: The deltas on which Bangladesh is built are subsiding, meaning the area will be more susceptible to the flooding that will accompany sea level rise.

Subsidence is a natural feature of deltas. The water in a large river may reach the sea in one or more branches, called distributaries. As the distributaries evolve, they may meander more and more, thus lengthening the river's route to the sea. During floods, the distributary can overtop its banks and cut a new and shorter route to the sea. This can result in a rapid shift in where sedimentation occurs at the delta's mouth. An existing delta can sometimes be quickly abandoned, and without the constant addition of new sediment, it subsides.

Here's Dr. Begum again: "A one-meter (3.3 ft) rise in sea level could turn a moderate storm into a catastrophic one. The tropical cyclones of the Bay of Bengal form only when sea surface temperatures reach 27 degrees C or higher. Tropical cyclones in the Bay of Bengal generally form in the pre-monsoon (April–June), and post monsoon (October–December) seasons. An increase in the surface air temperature (due to global warming) might lead to a more widespread occurrence of cyclones in the Bay of Bengal."

Clearly Bangladesh faces the potential for catastrophe from the combination of coastal population growth, coastal subsidence, land-use changes in the catchment area, and global climate change. As you know, global climate change is thought to be strongly influenced by the release of vast quantities of greenhouse gases, mainly from the burning of fossil fuels in industrial countries and China. Bangladesh's 1992 emissions of CO_2, the principal greenhouse gas, totaled 17. 2 *million* metric tons (37.8×10^9 lb), compared to the Untied States' 4.9 *billion* metric tons (10.8×10^{12} lb), which was by far the world's major source, more than double that of the former Soviet Union. The per capita emissions were even more disparate: 0.15 metric tons (330 lb) per Bangladeshi compared to 19.13 metric tons (42,086 lb) for each resident of the United States.

Now go to our web site to continue the analysis.

Find the complete issue discussion and project at www.prenhall.com/oceanissues.

[1] Bryant, Dirk, et al., 1995. Coastlines at Risk: An Index of Potential Development-Related Threats to Coastal Ecosystems. World Resources Institute (WRI) Indicator Brief. WRI, Washington, D.C.

[2] Begum, S. 1996. Climate Change and Sea-Level Rise: Its Implications in the Coastal Zone of Bangladesh. *In* Global Change, Local Challenge: HDP (Human Dimensions of Global Environmental Change Programme) Third Scientific Symposium 20–22 September 1995. Geneva.

[3] Mazhar, Farhad."What We Want From Kyoto". OTN explores Global Warming. Available: http://www.megastories.com/ warming/bangla/kyoto.htm. 6 April 1998.

[4] For details on storm surges go to the USGS home page at www.usgs.gov.

Catch of the Day: ~~Slimehead~~, Orange Roughy, ~~Patagonian Toothfish~~, Chilean Sea Bass

- What is the state of global fisheries?
- What are the trends in "harvesting" of wild marine fish?
- What is bycatch?
- What is the environmental impact of commercial fishing?
- How much seafood do we eat?
- How important is seafood as a protein source?
- Is "sustainable fisheries" an oxymoron?

Introduction

Wild Oats Markets, a nationwide chain of more than 75 grocery stores, issued a press release on August 11, 1999[1] stating that it would no longer sell North Atlantic swordfish, marlin, orange roughy, or Chilean sea bass because these species are endangered due to overfishing.

Paul Gingerich, Meat and Seafood Purchasing Director, commented, "Floods, drought and overgrazing that affect food sources on land can be readily seen and measured. The effects of overfishing cannot be easily seen in the oceans. We need to be proactive to save these species for future generations."

Will the other 246,000[2] U.S. grocery stores follow suit? Are the species in question, and others, really endangered?

The Impact of Global Fisheries

Even though less than 1% of global caloric intake comes from fish[3], the importance of fisheries to the global and many national economies cannot be overstated. Consider the following information on fisheries worldwide[4]:

- The value of the international fish trade for 1994 was $47 billion.
- The combined value of canned, fresh, and frozen fishery products in the U.S. in 1996 was over $2.9 billion.
- Nearly 85,000 people were employed in processing and wholesale jobs alone in the U.S. in 1995.
- The economies of many countries such as Iceland, Peru, and Norway depend heavily on fish product exports.
- In eastern Canada, the closure of the cod fishery cost at least 40,000 jobs, in a country with a population one-tenth that of the U.S.
- Of the $752 the average American spent on meat in 1995, $97 (13%) was for seafood (Figure 1).

Although the contribution of fish to the diet of humans may seem small if we simply count calories, it becomes critical if we consider protein: 16% of global animal protein is provided by fish, while in the Far East, where most humans

Figure 1

Examples of seafood marketing in the U.S. An all-you-can-eat seafood bar (top), featuring marine organisms from around the world, is a mainstay at many oceanside resorts. *(Photo Courtesy of Sunny Day Guide).* An Asian market in San Diego features moonfish (bottom), a thin, bony species often considered bycatch (see text). *(Photo by Abel).*

live, nearly 28% of animal protein comes from fish[5]. In developing countries worldwide, where population growth rates are ominously high, 950 million people depend on fish as their primary source of protein[6].

In addition to serving as a basic food source, fish is increasingly considered by affluent westerners to be a "health food." Fatty fish like salmon and mackerel have relatively high levels of omega-3 fatty acids, which have been shown in clinical studies to reduce the risk of heart attack by 50–70%.

Finally, fish and fish by-products representing as much as one third of wild-caught fish are a mainstay of the pet food industry and are used as a constituent of animal feed as well.

Environmental Costs of Fishing

Commercial fishing can be a very expensive as well as an environmentally costly activity. *First, many fishing methods destroy habitat.* Consider trawling, which is typically done for shrimp and other bottom-associated species like Atlantic cod and plaice. In this method, a 10 to 130 m (33–426 ft) long net is scraped across large areas of the bottom, collecting virtually everything in its path, including endangered sea turtles. Trawling, which became popular with the advent of the diesel engine in the 1920s, is practiced worldwide on virtually every different bottom type[7]. Saturation trawling, in which the net is repeatedly fished in an area until virtually no fish or shrimp are left, has been compared to clearcutting a forest. The comparison is appropriate: trawling heavily damages sessile benthic organisms like sponges, hydroids, and tube-dwelling worms and displaces associated fauna like fish and crustaceans. Complete recovery in both cases may take decades or longer. The comparison between trawling and forest clearcutting breaks down when one considers the area involved annually. Approximately 100,000 km^2 (38,000 mi^2) of forest are lost annually, whereas an area 150 times as large is trawled[8]!

Extremely destructive methods like dynamiting and poisoning still are used in some areas.

Commercial overfishing threatens fish stocks, which are already under stress from coastal environmental degradation due to overdevelopment and industrial, municipal, and agricultural pollution. A 1994 United Nations Food and Agricultural Organization (FAO) analysis of marine fish resources concluded that 35% of 200 top marine fisheries were overexploited (i.e., yielded declining landings); 25% were mature (and thus on the verge of endangerment if stressed); and 40% (largely in the Indian Ocean) were still developing.

Significantly, *no* major fisheries, according to the study, were undeveloped[9].

Additionally, *at least 25% of the commercial catch is unused.* This quantity is known as "bycatch" and refers to undersized, low-value, and non-target species (fish, crabs, etc.). Bycatch is often returned to the water dead, or dies soon after (Figure 2). The fishing activity that probably generates the most bycatch is trawling for shrimp. In addition to damaging the ocean bottom, as much as 90% of the trawl contents may be non-target and hence unused species, sometimes called "trash fish" by fishers.

Figure 2

Bycatch includes sea turtles, mammals, birds, invertebrates, and fish, like this oceanic sunfish caught in a Japanese driftnet in the Tasman Sea. *(Photo Courtesy of Greenpeace, Inc.)*

Innovations like TEDs (Turtle Excluder Devices) shunt large objects like sea turtles out of the net. The use of TEDs, however, is not universal and is not entirely successful either.

Finally, *the environmental cost of commercial fishing extends to the pollution* associated with the manufacture, transportation, and use of equipment (like fishing boats) and supplies; fuel spills; and transportation and refrigeration of fishery products.

Case Study: The Slimehead and Patagonian Toothfish

Would you eat a fish called a slimehead? Or a Patagonian toothfish? Probably not, so clever marketing specialists transformed these into popular items by renaming them as "orange roughy" (Figure 3) and "Chilean sea bass" (Figure 4). The former was popular in the mid to late 1980s, whereas the latter's popularity is just peaking.

Unfortunately, renaming the fish changed neither their biology nor their fate. The orange roughy is a classic example of ignoring the importance of gaining a complete understanding of a species' biology before exploiting it as a fishery. The Chilean sea bass is yet another reminder that we refuse to learn from our mistakes. Both cases demonstrate the power and environmental destructiveness of effective marketing.

The orange roughy fishery began off New Zealand in 1978 and quickly exceeded 35,000 tons (31.8 × 10^6 kg). Unfortunately, because the species is long-lived (how ironic!), reaches sexual maturity late in life, and doesn't produce profuse numbers of offspring, by 1990 the harvest had been reduced over 70%. Maximum sustainable yield, the amount fisheries managers estimate can be annually harvested without damaging the population, had been estimated to be 7,500 tons (6.8 × 10^6 kg), which is still lower than the 1990 harvest.

Chilean sea bass have a similar biology, and thus are likewise vulnerable to overfishing by powerful modern fishing methods.

Figure 3

The erstwhile slimehead, successfully marketed as the orange roughy. *(Jean-Paul Ferrero/Jacana, Photo Researchers, Inc.)*

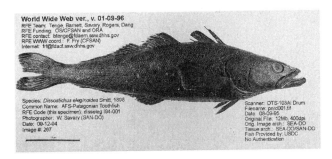

Figure 4

The fish formerly known as the Patagonian toothfish, now available in Western restaurants and seafood shops as the Chilean sea bass. It is not at all closely related to the group of fishes commonly known as sea basses. *(W. Savary, Center for Food Safety and Applied Nutrition)*

Stop now and search the web using the key words "Chilean sea bass." Did you find references on its biology, fishery, and conservation or did your "hits" contain mostly recipes?

Aquaculture (Fish-Farming)

Although "capture" fisheries are nearly fully exploited, and in many cases are dangerously overexploited, total fish production is increasing. This increase is a direct result of growth of aquaculture, mainly in China. From 1984–1993, world aquaculture production rose from 6.3 to 14.5 million metric tons (6.3×10^9 to 14.5×10^9 kg), while in China it increased from 2.2 to 8.1 million metric tons (2.2×10^9 to 8.1×10^9 kg) over the same interval[10].

"Finfish" such as tilapia, salmon, carp, and flounder comprise about one half of aquaculture production. Molluscs, mainly oysters and mussels, account for one fourth. Two thirds of aquaculture takes place in inland rivers, lakes, ponds, and artificial tanks. Coastal marine aquaculture (mar-

iculture), which includes the margins of bays, in bays and in the open ocean, accounts for the remainder.

Aquacultural output, like capture-fisheries production, is used for both fishmeal and canned, frozen, and fresh products. One variety, shrimp mariculture, alone grew by 350% between 1984 and 1993, in response to market demands and often to government subsidies[11].

Environmental Impact of Aquaculture

The greatest problems posed by marine fish-farming relate to land-use changes[12]. Tens of thousands of hectares of salt-water wetlands, mainly tropical mangrove[13] environments, are lost annually by conversion to shrimp mariculture alone. Mangroves are environments critical for the health of offshore coral reefs, since the roots of the mangroves trap sediment from rivers and creeks which could otherwise smother the reefs, and reefs, in turn, are important fishing grounds for many subsistence fishing communities, especially in Southeast Asia. Mangroves are also important nursery grounds for marine fish. Degradation of water quality and changes in benthic invertebrate fauna may also be caused by aquaculture.

To find out more about global fisheries and endangered fish stocks, go to our website.

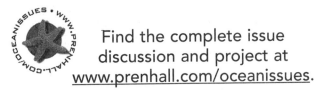

Find the complete issue discussion and project at **www.prenhall.com/oceanissues**.

[1] Available at www.wildoats.com

[2] *Statistical Abstract of the U.S. 1996-97*, Table No. 1283, p. 770

[3] Pimentel, D. and M. Pimentel (eds.) 1996. Food, energy, and society. University Press of Colorado, Niwat, Co.

[4] This information was obtained from a variety of sources, including the National Marine Fisheries Service, Statistical Abstract of the U.S. 1988, and World Resources 1996-97.

[5] Food and Agricultural Organization of the United Nations. 1993. Marine fisheries and the law of the sea: a decade of change. FAO Fisheries Circular No. 853. Rome.

[6] World Resources Institute. 1996. World Resources 1996-97: a Guide to the Global Environment. Oxford University Press, N.Y.

[7] Waitling, L. and E.A. Norse 1998. Disturbance of the seabed by mobile fishing gear: a comparison to forest clearcutting. Conserv. Biol. 12: 1180-97.

[8] Ibid.

[9] Food and Agricultural Organization of the Untied Nations. 1996. Chronicles of Marine Fisheries Landings (1950-1994): Trend Analysis and Fisheries Potential. FAO Fisheries Technical Paper No. 359, Rome.

[10] Worldwatch Institute. 1996. Vital Signs. W.W. Norton and Co., New York.

[11] Malcolm C., M. Beveridge, L.G. Ross, and L.A. Kelly. 1994. Aquaculture and biodiversity. Ambio, Dec. 1994.

[12] The weakening of genetic diversity of fish stocks poses a problem as well.

[13] Mangroves are salt-tolerant trees found worldwide in the tropics which function as coastline stabilizers, nursery grounds, and habitats for marine organisms.

Coral Rocks!
The Value of the World's Coral Reefs

- What are coral reefs, where are they located, and what organisms contribute to reef construction and function?
- What is the importance of coral reefs as reservoirs of biodiversity?
- How important are coral reefs to marine fisheries?
- What is their significance in the carbon cycle?
- What are the global threats to the health of coral reefs?

Introduction

Coral reefs (Figure 1) are tropical, shallow-water, limestone mounds formed mainly by coral animals and plants which remove calcium carbonate ($CaCO_3$) from seawater and deposit it as skeletal material, a process known as "carbonate-fixing."

Coral refers to marine invertebrate organisms belonging to the Phylum Cnidaria, Class Anthozoa, or to the hard, calcareous structures made by these organisms. An individual coral animal, known as a polyp, resembles a minute anemone. It is the ability of these polyps to remove dissolved calcium carbonate from seawater and to deposit it as part of their rocky skeleton that allows the formation of coral reefs. How coral animals do this is not precisely known. What *is* known is that corals need algae actually living in their tissues in order to precipitate the $CaCO_3$. (More about this below.) This type of relationship, termed symbiosis, or more specifically, mutualism, occurs among a wide variety of marine organisms.

Coral reefs represent some of the most important real estate on the planet. They cover approximately 600,000 km^2 (230,000 mi^2) in the tropics, an area roughly comparable to the state of Texas. They are a major oceanic storehouse of carbon and may contain up to a million species of organisms, only a tenth of which have been identified.

Coral reefs are increasingly under threat from human actions. For example, nearly half a billion people live within 100 km (62 mi) of coral reefs. This issue focuses on the importance of coral reefs, their role in global carbon balance, and what can be done to protect them.

Figure 1

Satellite photograph of the Great Barrier Reef. The Great Barrier Reef is the largest structure built by living organisms. It stretches along Australia's east coast for 2,000 km (1,200 miles). Like most organisms, corals live near the maximum temperature they can tolerate. Will rising sea temperatures result in the disappearance of corals? *(NASA/Johnson Space Center)*

Background

While reef environments have been fairly common in Earth's history, and reefs have been built by organisms on and off for the past several billion years, not all reef structures have been built by corals. Corals have been important contributors to reefs only since about 300 million years ago, and only within the past 50 million years have modern corals assumed their reef-building roles. Noncoral reefs built by cyanobacteria were accumulating more than 3 billion years before present, and thus are among the most ancient structures built by organisms.

Along with tropical rainforests, coral reefs, particularly those in the tropical Indo-Pacific Ocean, have the highest known biodiversities of any ecosystem on earth. But globally, coral reefs are not prospering; indeed, their very existence is being threatened by nutrient pollution, sedimentation, overfishing, global warming, and even ecotourism. The severity of this problem can be appreciated if you realize that tourists spend upwards of $100 billion each year to visit locations near reefs. Florida reefs, for example, bring in at least $1.66 billion annually to that state's economy.

Coral Reefs and Fishing

Although reefs cover less than 0.2% of the ocean surface, they harbor a quarter of all marine fish species. Fish in coral communities are of two basic types—herbivores that feed on algae and carnivores that eat other animals.

Fishers have been plying reefs for millennia, and today reefs provide employment for millions of fishers. During the past few decades, however, the intensity of fishing has begun to degrade reef communities. Damage to Philippine reefs has resulted in the loss of 125,000 jobs. You may be able to anticipate some of the reasons for the degradation of coral reefs: if too many herbivores are removed, the marine algae that they eat may grow out-of-control and smother the coral (Figure 2). Removing carnivorous fish can also upset the reef's ecological balance. Can you see why?

While removal of the fish may harm reefs ecologically, some methods of collection, such as dynamiting, kill the hard coral colonies directly. Another method, called muro-ami, involves bouncing rocks tethered to lines off the coral to herd the fish. This method, employed in the Philippines, typically destroys about 17 m^2 (183 ft^2) of coral cover per hectare (10,000 m^2 = 108,000 ft^2) per operation. Typically thirty muro-ami boats repeat the process about ten times a day. It may take forty or more years for reefs that are destroyed by fishing practices to recover (if allowed to).

Macroscopic marine algae grow amid and on the coral. Some of these algae are themselves carbonate producers and may be second only to corals in their carbonate production. These algae, as well as noncalcareous algal species, may be eaten by parrotfish and other herbivores, who find safety in the numerous nooks and crannies of the reef itself. These fish in turn become food for predators.

Figure 2

Algae overgrowing staghorn coral in the Caribbean.
(© Stephen Frink/CORBIS)

Reef Health

That coral reefs are widely distributed in the tropics does not imply that the coral animals themselves are very hardy. In fact, quite the opposite is true; corals are sensitive to such a variety of environmental factors that it is a won-

der they exist at all. These factors are light, temperature, salinity, sedimentation, and nutrient levels.

Light Reef-building (also called *hermatypic*) corals are restricted to shallow waters because they require a certain quality and quantity of light. Why would simple, eyeless, invertebrate animals such as corals have a need for light? This requirement is due to the presence of simple, one-celled algae known as zooxanthellae which live within coral cells.

Zooxanthellae are dinoflagellates, a group that also includes the organism responsible for red tides. These organisms, thousands of which live within the cells of a polyp, require light to carry on photosynthesis. Sugars, the high-energy end-products of photosynthesis are used to nourish not only the zooxanthellae but the coral as well. Some of the organic products are transferred to the coral polyp as a nutritional supplement to the food obtained when the coral feeds using its tentacles. Zooxanthellae also contribute oxygen and remove some waste materials, produced by the polyp and they are involved in calcification.

Temperature In addition to light, corals require a relatively constant, moderately high (but not too high) temperature. The global distribution of coral, in fact, correlates best with surface temperature: corals are not generally found where winter surface water temperatures fall below 20°C, and they generally expel their algae and may die if water temperatures exceed 30°C. Thus, reefs are generally confined to waters between the Tropics of Cancer and Capricorn. This location protects them from low water temperatures, but means that most corals live near their upper thermal limit of tolerance.

Salinity cannot vary much from the 35 part per thousand average for seawater for corals to survive (except for Red Sea corals, which have become adapted to higher salinities).

Sedimentation rates should be low and grain size relatively coarse or corals can be easily smothered. First, the corals' filtering apparatus can be clogged, making it difficult for them to feed on the tiny creatures that comprise their diet, and second, the sediment can blanket the colony and keep the zooxanthellae from photosynthesizing. Sedimentation due to runoff from construction sites onshore or mining activities is one of the reefs' most lethal enemies.

Nutrient levels (that is, the concentrations of phosphorus and nitrogen compounds) must also be low. In fact, most coral reefs flourish in nutrient "deserts." The reason again is fairly easy to understand. High nutrient concentrations may stimulate the growth of algae, which can smother the corals. Expansion of intensive, western-style agriculture on land (with its massive doses of water-soluble, nutrient-rich fertilizer) may lead to die-offs in offshore reefs. And in the United States, treated, nutrient-rich sewage from Florida's

West Coast may be among sources of degradation affecting the once-healthy coral reefs in the Florida Keys.

Reefs and the Carbon Cycle

Coral reefs are an integral part of the planet's carbon cycle, which may ultimately help control the earth's surface temperature range. The living polyps of a coral reef represent only a thin film covering a massive rock skeleton built up over centuries or millennia by generations of coral polyps. This hard foundation is made of calcium carbonate, which is derived from calcium ions and carbon dioxide dissolved in seawater. As the coral polyps grow, they (along with calcareous algae) precipitate calcium carbonate (that is, convert it from a dissolved form to a solid form), which supports the growing colony of coral and helps to cement coral rubble together. In the process, carbon dioxide is removed from the water. This carbon dioxide "deficit" in the water is replenished from the CO_2 in the atmosphere.

Corals are extremely efficient at fixing carbon (remember, we are talking about the carbon cycle, and carbon is a key component of calcium *carbonate*). Each kilogram of $CaCO_3$ contains almost 450 grams (1 lb) of carbon dioxide.

(You can see why if you figure out the atomic weight of $CaCO_3$, and then determine the proportion of this that is CO_2. Stop and do this now.)

Geologists, by the way, believe that the earth's early atmosphere was greatly enriched in carbon dioxide, mainly from volcanic gases. As life evolved, some marine bacteria began to precipitate calcium carbonate, in effect storing or fixing carbon dioxide, removing it from the atmosphere. As the CO_2 was gradually removed from the early atmosphere and stored in rocks, the ability of the atmosphere to absorb heat diminished.

Today, immense volumes of carbonate sedimentary rock on continents attest to the enormous amount of carbon dioxide removed from the atmosphere. If this CO_2 were restored to the atmosphere, the earth's surface temperature might approach that of Venus, where surface temperatures (460°C) are hot enough to melt lead.

Figure 3

Brain coral in the Caribbean showing extensive bleaching. (© Stephen Frink/CORBIS)

Coral Bleaching

Thus, for a variety of reasons, coral reefs should be protected and their growth encouraged. Yet, in 1998, scientists documented one of the severest threats to coral survival in recorded history. The threat is called "bleaching" and it happens when corals expel the zooxanthellae living in their tissue (Figure 3). The greatest stress seems to come from high water temperature, which would be a consequence of global warming. Data from the past century suggest that such bleaching events could become more severe and persistent as the planet warms. This global event compounds the stresses on reefs from local human-induced sources that we described above. Exposure of coral to increased UV radiation due to the seasonal thinning of the ozone layer may be yet another significant problem.

What is the future of coral reefs? Unless humans take these threats seriously and act together to mitigate them, the future is not bright.

Find the complete issue discussion and project at www.prenhall.com/oceanissues.

Global Warming and Sea Level Rise

- What factors cause changes in sea level?
- How can scientists estimate the magnitude of sea level rise given the thermal expansion of seawater and melting of ice associated with sustained global warming?
- How can we assess the severity of impact of sea level rise to shallow coastal communities and human populations given the best estimate of sea level rise over the next 100 years?

Introduction

2:06 am 02/22/2005; You wake up abruptly as police sirens begin blaring outside of your home. As you stumble to turn on the light, you hear something being said on a bullhorn about evacuation. A minute later, air raid sirens join in the frenzy. You turn on the television to catch the news, and there are images of Antarctica, and ice, and enhanced satellite images. What could possibly be the connection between the peaceful and serene images of the frozen continent and all of this noise? You listen carefully to the live news report. "At 11:53 PM yesterday evening—slightly more than two hours ago, a large portion of the West Antarctic Ice sheet slid violently into the adjoining polar ocean." You think, "Well that's great news, but what does it have to do with me, and why all of this fuss?" The news anchor continues, "All residents of coastal states living within 200 miles of the ocean or coastal bays are requested to evacuate immediately. The National Oceanographic and Atmospheric Administration has predicted that the water displacement generated by the slide of this ice sheet into the ocean will send a tsunami wave of 10 meters along the edge of the Atlantic Ocean basin. It is predicted that the South Atlantic regions of Florida, Georgia, and South Carolina will be impacted by this wave in roughly 16 to 17 hours. This evacuation will be permanent because the global sea level after the wave passes will be 6 meters above current sea level. We now transfer you to the press room of the White House for an emergency address by the President." You pinch yourself hard on the cheek thinking this must be a dream, this is impossible! But is it?

Background

It is thought that water began to accumulate on Earth's surface between 4.2 and 4.4 billion years ago, after the crust cooled below the boiling point of water. Water that now covers 71% of the earth's surface was derived from two main sources, outgassing of water-laden volcanic gases, and from fragments of comets impacting the upper atmosphere. Compared to the Earth's early history, the rates of input of water onto the earth's surface has slowed, and it is now thought that the volume of water (in all its states) on the surface has been relatively constant for several billion years. This implies that there are also mechanisms causing a slow loss of water away from Earth's surface. These losses could be due to recycling of Earth's crust and water-laden ocean sediments, as well as the slow escape of hydrogen resulting from the disassociation of water molecules from the upper atmosphere into space.

This relative constancy of water mass on the planet over the last several billion years does not, however, mean that sea level has remained constant. Still, the 220-meter range of sea level variation over this period is small compared to the mean depth of the ocean–3,800 meters.

Sea Level Changes

Changes in sea level, which have occurred over most of the earth's geologic history, are due to two processes. *Eustatic* processes change the absolute amount of liquid water within the ocean basins. The main mechanism driving eustatic changes in sea level is the reproportioning of water between liquid and solid phases (ice) due to changes in global climate. *Isostatic* processes change the underlying topography of the sea floor. These changes can occur on either regional scales, as in the rebound of crust after deglaciation, or in the slow subsidence of deltas at passive continental margins, or on global scales. During periods of marked increases in sea floor spreading rates the height of ridge and rise features increases throughout the global ocean basin, in turn causing displacement of water upward onto the coastal continental landscape.

Changes in sea level have been implicated directly and indirectly as contributing to mass extinction events that have occurred within the earth's geologic past. Rapid sea level decline has even been hypothesized to cause changes in atmospheric oxygen levels. In this case, rapid decay (oxidation) of shallow, newly exposed organic-rich marine deposits would remove oxygen from the atmosphere.

Sea level has been rising since the end of the last glaciation about 15,000 years ago. The rate of global sea level rise for the last 100 years has been 2 mm/year and 15–20 cm total (0.08 in/year, 6–8 in. total). Estimates for global sea level rise for the next century suggest that this rate will double to 4 mm/year and 40–45 cm total (0.16 in/year, 16–18 in. total).

The observed rate of coastal sea level rise varies from region to region because of the variability in isostatic crustal movement (Figure 1). If future increases in sea level become more rapid, shallow water intertidal or subtidal communities may not be able to keep pace with sea level change and will die off as environmental conditions change beyond their limits of tolerance. Such a change is now occurring in intertidal salt marsh environments around the Mississippi River delta in Louisiana, as the coastal landscape subsidence combined with global sea level rise exceeds the rate of vertical growth of the marsh community.

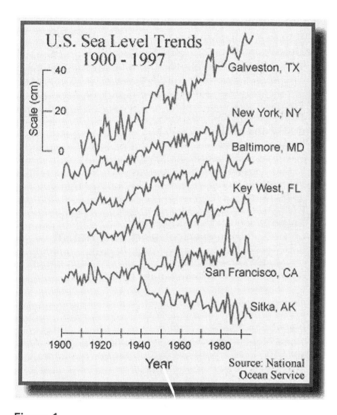

Figure 1

Trends of sea level rise within coastal city regions of the United States. Note the higher rate of sea level rise for Galveston, Texas which is experiencing coastal subsidence, whereas in Sitka, Alaska sea level appears to be falling because of coastal uplift. (*NOAA National Ocean Service*)

Impacts of Global Warming

Although ample scientific evidence now exists of geo-historical impacts of changing sea level upon biological communities, humans have only recently become concerned with the potential of such processes to change their lifestyles and standards of living. This newfound concern stems from the virtual consensus among climatologists that the planet is experiencing a period of global warming.

Global warming may result in an increase in the rate of sea level rise due to thermal expansion of seawater and the melting of ice in glaciers and polar regions (Figure 2). The coefficient of thermal expansion of seawater is 0.00019 per degree Celsius, meaning that if a volume of seawater occupied 1 cubic meter of water (1000 L, 264.20 gal), after warming by 1°C, it would expand to 1.00019 m³ (1000.19 L, 264.25 gal). Translated over the mean depth of the ocean (3.8 km, 2.4 mi), an increase in temperature of 1°C will cause a sea level rise of about 70 cm (28 in). Whereas thermal expansion acts upon water already in the basin, contributions from melting ice represent new, added water to the present ocean volume. The melting of ice that is currently perched upon land, as in the ice sheets of Greenland, Iceland, and the Antarctic, has the potential to raise sea level considerably, by about 80 meters (262 ft). Ice that is already floating in the ocean water, as in the Arctic ice mass, Antarctic ice shelves, and much smaller icebergs, may melt but will not contribute to sea level rise since the mass of water contained in these features already displaces its equivalent water volume.

There is ample evidence that the increase in melting has begun. Monitoring of mountain glaciers throughout the world over the last two decades indicates that 75% of them are losing mass and that the rate of this loss has almost doubled, from 0.25 m to 0.50 m (0.82 to 1.64 ft) of water equivalent, in the last 20 years. In 1991, NASA reported that the extent of sea ice in the Arctic Ocean declined by 2% between 1978 and 1987. More recently, Norwegian scientist Ola Johannessen, of the Nansen Environmental and Remote Sensing Center, has presented satellite measurements of microwave emissions that show declines in permanent Arctic ice of 7% over each of the past two decades.

Figure 2

Landsat 1 MSS digitally enhanced image of the Byrd Glacier in Antarctica, where it joins the Ross Ice Shelf. Melting of glaciated areas will directly increase sea level, whereas the melting of shelf ice will not cause a rise in sea level but may be an important indicator of regional climate change. (*NOAA Central Library Photo Collection*)

In the Antarctic, a series of ice shelves (Wordie, Larsen a, and Larsen b) have collapsed over the last decade spurred by the 2.5°C increase in the average temperature on the peninsula since the mid-1940s. There is some concern that loss of these ice shelves may destabilize continental ice sheets, with the Western Antarctic Ice Sheet being considered as potentially capable of slippage into the surrounding ocean.

Which Areas Will be Impacted

Sea level rise will produce increased flooding risks in areas already under sea level, like New Orleans, Louisiana in the United States and coastal Holland in Europe (Figure 3). The impact of sea level rise will also be severe for certain small island nations. A series of low lying islands in the Pacific (Marshall, Kiribati, Tuvalu, Tonga, Line, Micronesia, Cook), Atlantic (Antiqua, Nevis), and Indian Oceans (Maldives), will be greatly impacted. For example, in the Maldives most of the land is less than 1 meter (3.3 ft) above sea level. A seawall recently built to surround the Maldivian capital atoll of Malè cost the equivalent of 20 years of the entire Maldivian gross national product, according to U.N. reports. Coastal regions that possess little geographic relief and regions which are also experiencing subsidence due to sediment accumulation are also at threat.

The largest population at risk are the people living in Bangladesh. About 17 million people in Bangladesh live less than 1 meter above sea level. In Southeast Asia, a number of large cities, including Bangkok, Bombay, Calcutta, Dhaka, and Manila (each with populations greater than 5 million), are located on coastal lowlands or on river deltas. Particularly sensitive areas in the U.S. include the states of Florida and Louisiana, coastal cities, and inland cities bordering estuaries.

Recent technological advancements in climate research and satellite remote sensing have enabled scientists to more fully understand the mechanisms that cause climate change and sea level response. Paleoclimate tools include an arsenal of fossil, geophysical, and geochemical indicators that are sensitive to changing climate and can be interpreted back in time to provide evidence of past environmental conditions. Such data can then be applied to model the temporal and spatial patterns of future climate change and sea level response. Satellite remote sensing now permits the collection of large-scale regional and global data sets that prior to the 1950s were only a dream. Such an increase in the extent of examination allows earlier detection of climate change, as it integrates important areas, previously difficult to study, such as the extreme polar regions. Other useful information is coming from declassified military information on sea ice draft (thickness under water) from the Arctic region. Submarines and permanently moored devices are now continually monitoring this portion of the earth's ice mass.

Find the complete issue discussion and project at www.prenhall.com/oceanissues.

Figure 3

Flood control structures along Eastern Holland dwarf the small boat in the foreground. Investment of countries in the production of such structures represent a considerable proportion of the country's gross national product. (© Patrick Ward/CORBIS)

Human Impacts on Estuaries

- How are estuaries formed?
- What contributes to the high biological productivity in estuaries?
- Why are estuaries important to human society?
- Why are estuaries exceptionally prone to human impacts?

Introduction

Today's estuaries are all less than 12,000 years old, reflecting the geologically recent rise of sea level since the last glacial maximum. Estuaries are typically ephemeral coastal features with youthful ecosystems.

Ecological Function of Estuaries

The shallow, nutrient-rich waters of estuaries and associated wetlands create a highly productive environment for plants and animals. In fact, estuarine environments are among the most productive on earth. The high concentration of nutrients and shallow depth support phytoplankton, seagrasses, macroalgae, emergent grasses, and, in tropical environments, mangroves.

Animal communities in estuarine sediments are of low diversity, since many species cannot tolerate the extreme fluctuations in temperature and salinity. Animals that feed upon benthic (bottom-associated) organisms are in turn food for larger consumers, which include larger fishes as well as birds and mammals.

Many subhabitats within estuaries, such as salt marshes and seagrass beds, serve as nursery grounds for the juvenile stages of commercially valuable fish and shellfish. The production of the higher plants (seagrasses, marsh grasses, and mangroves) is in great surplus of what is directly consumed, and the decomposition of this material leads to a detrital food web where a large portion of the nutrition to the consumers derives from the microbes that grow on the dead plant material.

The wetlands that fringe many estuaries also perform valuable services for human society. Estuaries provide habitat for more than 75% of America's commercial fish catch and for 80–90% of the recreational fish catch. Water draining from the uplands carries sediments, nutrients, and pollutants. As the water flows through fresh and salt marshes, much of the sediments and pollutants is filtered out. Wetland plants and soils also act as a natural buffer between the land and ocean, absorbing flood waters and dissipating storm surge. This protects upland organisms as well as valuable real estate from storm and flood damage.

Population Growth in Coastal Regions

Features of estuaries which favor their high productivity, namely their semi-enclosed nature, riverine inflow of nutrients, and extensive intertidal plant communities, also attract humans. Many major cities are located along estuaries (Boston, Baltimore, Charleston, Norfolk, San Francisco, Seattle, etc.). Historically, these cities were, or are still, major ports for commercial shipping.

River connection to estuaries expanded the ability of cities far from the estuary mouth to develop port operations and dependent industry. The floodplain region of estuaries and rivers possesses rich soils that enhance agricultural production, and having nearby ports for marketing these products encouraged agriculture throughout the watershed region of the estuary.

The aesthetics of rivers, estuaries, marshes, and beaches have also encouraged a dramatic growth both of residents and tourists. It is therefore not surprising that population growth has been much greater in coastal and estuarine regions of the U.S. compared to the inland regions (Figure 1). More than 139 million people—about

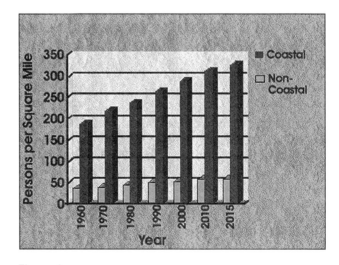

Figure 1

Population growth of coastal and non-coastal constituents of the U.S. from 1960 to estimated values in 2015. Data from the NOAA State of the Coast Report, 1998. *(NOAA)*

53% of the national total—reside along the narrow coastal fringes. This coastal population is increasing by 3,600 people per day. The increase in human population and their associated activities within estuarine watersheds have caused significant declines in water quality and changes in the structure and function of estuarine ecosystems.

Human Effects Upon Estuarine Water Quality

Human activities have altered estuarine watersheds throughout the world. Increases in nutrients (eutrophication) occur naturally in aquatic systems as they age, but human population growth has accelerated the accumulation of these nutrients many orders of magnitude faster than what would occur naturally. High rates of nutrient input into estuaries can contribute to fish disease, toxic and nontoxic algae blooms, low dissolved oxygen, and marked change in plankton and benthic community structure.

Toxicants and pathogens also enter estuaries. Toxicants, including organic substances and metals, may exert direct and acute effects upon the aquatic community, or they may be bioconcentrated as these materials are passed up the food web, which selectively impacts the higher trophic level species. Exposure to toxicants or pathogens in coastal waters, or the consumption of undercooked or raw seafood harvested from those waters, can cause severe illness or death. Coastal population growth has led to an increasing flux of pathogens (viruses, bacteria, and parasites) to coastal waters, primarily from sewage.

The use of the fecal coliform test to judge the suitability of coastal waters for swimming and shellfish harvest has provided a significant level of protection (i.e., there have been few serious outbreaks of waterborne diseases), but illness and an occasional death still result from human pathogens in coastal waters. In the case of toxic phytoplankton, coastal monitoring occurs on a local basis, often by local health departments, as well as by the Food and Drug Administration.

Seafood safety is a significant concern to the nation. Paralytic, diarrhetic, neurotoxic, and amnesic shellfish poisonings are all caused by biotoxins accumulated from algae. Outbreaks of poisoning due to various toxins accumulated by shellfish and fish such as the recent *Pfiesteria* outbreaks in Chesapeake Bay have occurred several times in the past few years (Figure 2).

Some human activities directly modify the habitat of resident estuarine species. These include: conversion of open land and forest for commercial development, agriculture, forestry, highway construction, marinas, diking, dredging and filling, damming, and bulkheading. A good example of this problem is illustrated by the loss of prime spawning grounds for a variety of Pacific salmon species. The timber harvesting operations in the U.S. Pacific Northwest include expansive road construction and forest clearcutting. These activities generate a silt-like runoff that can smother the eggs of salmon in the upper reaches of streams in their natal watersheds. The precipitous decline of a variety of the west coast salmon is probably due, in part, to this habitat loss.

Changes in the hydrology of watersheds draining to the coast have also occurred as a result of landscape changes, channelization and damming, fresh water uses, and diversion to other drainage basins. Reductions in freshwater flow due to increased use or diversion of fresh water have caused problems especially in the southwestern coastal region of the United States.

Another example of this type of problem occurs with dredging. Dredging channels to promote port operations enhances the transport of salty oceanic water into bays and estuaries, changing the salinity structure, circulation, flushing, and water residence times of these semi-enclosed coastal systems. Such changes can have dramatic effects on biological productivity and ecosystem structure and function.

The exploitation of living and nonliving resources can affect coastal ecosystem health. Fishing provides many benefits to society, including food, employment, business opportunities, and recreation. However, like many human activities, fishing also can have deleterious ecological effects. Overharvesting of fishery resources may also create environmental problems. In Chesapeake Bay, the Atlantic oyster (*Crassostrea virginica*) used to cover extensive regions of the subtidal bottom, being capable of filtering the entire bay water volume over several days. The over harvesting of oysters during the last century has now eliminated commercial oyster reefs. The organic matter that used to be filtered by oysters now falls to the bottom and is consumed primarily by benthic microbes, leading to increasing frequency, extent, and duration of low oxygen (anoxia and hypoxia) conditions.

Figure 2

Sores on menhadden associated with an outbreak of *Pfiesteria piscicida*. The incidence of harmful algae blooms in coastal and estuarine environments have been increasing world-wide in the last several decades. *(J. Burkholder)*

In some coastal environments, foreign "exotic" species have been introduced by human activities and have established populations that have had major ecological consequences. The zebra mussel (*Dreissena polymorpha*) has settled into the Great Lakes and is now continuing to expand southward into fresh water regions of drainage basins into the Chesapeake Bay. In San Francisco Bay, the major dominant benthic organism is the nonindigenous Chinese clam (*Potamocorbula amurensis*), and its filter-feeding activities have eliminated the normal summer phytoplankton blooms in the northern portions of the bay.

Diseases that are further ravaging populations of oysters in Chesapeake Bay may have been introduced with oysters transplanted from other regions. Likewise, organisms transported for aquaculture and recreational fishing purposes in the past have been the source of many species introductions. More recently, the veined rapa whelk (*Rapana venosa*) has been found in Chesapeake Bay, where its feeding upon hard clams represents a significant threat to the last commercial shellfish resource currently being harvested.

Living near the coast, although highly desirable to most of our population, does carry additional threats of natural disasters like hurricanes (Fig. 3), storm erosion, and tsunamis. Great advances in weather forecasting technology now prevent the massive loss of life that previously occurred. However, loss of life still continues and property damage associated with larger coastal populations is exerting adverse effects on the insurance industry as well as the U.S. economy.

Beaches and wetlands surrounding estuaries buffer coastal land and habitats from assault by the ocean, providing the most effective means of preventing coastal erosion and habitat destruction. Shoreline erosion and haz- ardous storms are affected in a complex manner by land-use decisions and climate change and, conversely, can greatly affect coastal environmental quality. Studies of global climate change and improvements in the predictability of climate variability are crucial for predicting and mitigating the impacts of shoreline erosion and hazardous storms.

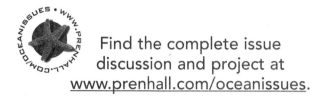

Find the complete issue discussion and project at www.prenhall.com/oceanissues.

Figure 3

Hundreds of pigs fight for their lives as flooding engulfs a large hog farm in North Carolina. Upwards of 100,000 hogs drowned as a result of Hurricane Floyd. (*The News and Observer/Chris Seward*)

Illegal Immigration:
Ballast Water and Exotic Species

- What are exotic species?
- How are they able to cross oceans?
- What enables them to colonize foreign environments?
- Do they represent a threat to coastal ecosystems?
- What can policy makers do to protect estuaries from non-native species?

NEWS FLASH!

In September 1999, U.S. President Bill Clinton issued an executive order that directed the Departments of Agriculture, Interior, and Commerce, the Environmental Protection Agency, and the U.S. Coast Guard to develop an alien species management plan within 18 months to blunt the economic, ecological, and health impacts of invasive species.

Agriculture Secretary Dan Glickman promised "a unified, all-out battle against unwanted plants and animal pests." But senior administration officials acknowledged that the task poses difficulties and may succeed only through greater international cooperation.

Officials acknowledged that the United States also has species that cause adverse impacts when they are carried to other countries. Interior Secretary Bruce Babbitt said the long-term answer is to resolve these issues through international agreements that would benefit all countries.

Environmentalists, meanwhile, complain that the Clinton Administration has been slow in regulating ballast discharges from freighters—one of the major pathways for exotic aquatic organisms such as the Chinese mitten crab (annual economic cost unknown)[1]; green crab (annual economic cost $44 million); and Asian clam (annual economic cost, $1 billion!), which are threatening native marine life in San Francisco Bay and as far north as Washington State. Another such animal is the veined rapa whelk (Figure 1), which has been recently discovered in Chesapeake Bay.

Since the issue involves interstate and international commerce, individual states and counties cannot, under the U.S. Constitution, regulate ballast water. The West Coast invaders, says Linda Sheehan of the Center for Marine Conservation in San Francisco, are driving out native crabs and clams and threatening local oysters—even burrowing into and weakening flood control levees, which could potentially result in huge losses from property damage during floods.

A coalition of environmental groups and the Association of California Water Agencies asked the EPA to regulate freighter ballast water discharges under the Clean Water Act.

Introduction

Silently, almost imperceptibly, the planet's oceans, seas, estuaries, and lakes are being invaded by plants, animals, bacteria, and even viruses from distant climes. These organisms are called alien, exotic, or invasive species. Sometimes their impact is negligible, rarely is it beneficial, and often it borders on the disastrous. At a January 1999 meeting of the American Association for the Advancement of Science, Cornell University ecologist David Pimentel estimated the total cost of invasive species at $123 billion a year.

Activity 1: Stop and write down one or more ways in which alien species can have a harmful effect on an ecosystem into which they have been introduced.

Instead of remaining in an ecosystem in which all members have evolved and interacted over time, in the relative blink of an eye invasive species may be transported beyond their natural range into the presence of other organisms with which they will immediately begin to interact, and perhaps compete.

Once thrust into a new environment, an organism may face a whole new set of conditions. To survive, all living organisms must live long enough to bear offspring and thus ensure the future of their gene pool. The 'aim' of exotic species is not to take over an estuary or clog a factory's water pipes, but rather to simply survive and reproduce.

Scientists believe that most non-native organisms fail to survive in their new environment long enough to become established. And that's a very good thing. But occasionally the introduced organism finds its new home completely livable, sometimes even ideal. Successful invasive species usually share a similar set of characteristics, according to the U.S. Coast Guard:

- They are *hardy,* indicated by their surviving a trip inside a ship for perhaps thousands of miles.
- They are *aggressive,* with the capacity to outcompete native species.

Figure 1

The veined rapa whelk. Scientists at the Virginia Institute of Marine Science are tracking this invader from the Sea of Japan. As of February 15, 2000, 650 confirmed observations of this species have been recorded. U.S. East Coast estuaries have favorable temperatures and ample prey (bivalves) for the rapa whelk, but they lack predators. One novel but ironic way of managing the invasion is to develop a fishery for the whelks as a means of controlling their population. *(Juliana Harding, Molluscan Ecology Program, Virginia Institute of Marine Science)*

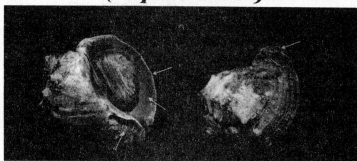

Have You Seen This Animal?
The Veined Rapa Whelk
(Rapana venosa)

The Virginia Institute of Marine Science (VIMS) is interested in any sightings of this large snail in Virginia waters. The veined rapa whelk is native to the Sea of Japan, reaching sizes of 5 to 7 inches in length. There are several distinguishing characteristics that are highlighted by arrows in the above pictures. Note the small teeth along the edge of the shell and the orange coloration along the inner edge of the shell. Other characteristic features are a pronounced channel (columella) and the ribbing at the lower end of the shell.

- They are *prolific breeders,* and can take quick advantage of any new opportunity, and
- They *disperse rapidly.*

Rapid dispersal is facilitated by having a planktonic larval stage, which allows the juveniles to be carried far and wide by currents. Such an introduced species often spreads rapidly, especially when predators and pathogens normally encountered in its home range are absent from the new environment, or when they are better able to feed than their new neighbors. (Or if they find their new neighbors especially tasty!)

In the above scenario, alien species flourish and potentially can reach astonishingly high population levels. Often, native species are displaced, or 'outcompeted' by the invaders. Then the situation is often called an *invasion.* Invasive species can inflict damage on ecosystems by:

- outcompeting native species,
- introducing parasites and/or diseases
- preying on native species, and/or
- dramatically altering habitat[2], e.g., rearranging the spatial structure of an ecosystem

Most invasive species are brought to new shores in the ballast water of ships, but animals dumped into an estuary from aquariums or accidental releases from aquaculture facilities may also contribute.

What is ballast water? Ballast water is carried by ships in special tanks to provide stability, optimal steering, and efficient propulsion. According to the U. S. Coast Guard, the use of ballast water varies among vessel types, among port systems, and with cargo and sea conditions.

How much ballast water is involved? The National Oceanographic and Atmospheric Administration (NOAA) has calculated that 40,000 gallons (150,000 liters) of foreign ballast water are dumped into U.S. harbors each *minute.*

The problem with ballast water is very simply stated: ballast water is taken up by a ship in ports and other coastal regions, in which the waters may be rich in planktonic (small, floating, or weakly swimming) organisms. It may be released at sea, in a lake or a river, or in the open ocean along coastlines—wherever the ship reaches a new port. As a result, a myriad of organisms is transported and released around the world within the ballast water of ships.

Here are two examples. Scientists studying an Oregon bay counted 367 types of organisms released from ballast water of ships arriving from Japan over a four-hour period[3]! Another study documented a total of 103 aquatic species introduced to or within the United States by ballast water and/or other mechanisms, including 74 foreign species[4].

Invasive Species and Chesapeake Bay

There is growing concern about invasive species' impact on Chesapeake Bay. How much of a potential problem can be assessed by these 1995 statistics from the Chesapeake Bay Commission:

- More than 90% of vessels arriving at Chesapeake Bay ports carried live organisms in ballast water, including, but not limited to, barnacles, clams, mussels, copepods, diatoms, and juvenile fish.
- Nonindigenous species have been responsible for paralytic shellfish poisoning, declining commercial and sport fisheries, and even cholera outbreaks!
- The two ports of Baltimore and Norfolk alone receive 2,834,000, and 9,325,000 metric tons of ballast water, respectively, each year, and this water originates from nearly fifty different foreign ports!

A database of organisms nonindigenous to Chesapeake Bay, prepared by the Marine Invasions Research Lab of the Smithsonian Environmental Research Center, lists 160 species and classifies another 42 as of uncertain origin.

Today, ballast water appears to be the most important means by which marine species are transferred throughout the world. As you saw above, the transfer of organisms in ballast water has resulted in the unintentional introduction of hundreds of freshwater and marine species to the U.S. and elsewhere.

Furthermore, the rate of new invasions from ballast water has increased in recent years[5]. To the extent these unwelcome visitors do economic damage, they make up a generally hidden cost of world trade.

Invasive Species Along the Pacific Coast of the U.S.

Chinese mitten crabs, another invasive species affecting San Francisco Bay and the Sacramento River Delta, are described by some scientists as "burrowing fiends," digging burrows that can significantly weaken levees (embankments built to prevent flooding) in a region which is prone to dangerous floods.

Activity 2: If the costs incurred during a flood are in part due to ships involved in international trade, how can these costs be fairly apportioned? Is it fair for only those people who are affected by floods in California to pay for the hidden costs incurred as a result of invasive species? Is it reasonable to price imported goods cheaply and then expect local residents to bear the cost of flooding resulting from this trade?

Suggest some possible solutions to this problem, but remember, localities and states do not have the right under the Constitution to regulate international trade.

Another growing problem is posed by the European Green Crab—a recent import that is affecting coastal California and seems to be working its way up the coast to Oregon and Washington. This aggressive predator prefers clams to oysters but could prey on baby Dungeness crabs (an economically important species) and smaller shore crabs.

And Washington State's oyster farmers are already uneasily coexisting with another exotic: the oyster drill that came from Japan with Pacific oysters. Already Washington oyster growers have had to abandon habitat overrun by the oyster drill.

By now you should have seen ways that invasive species can materially affect the American economy as well as our environment. Now go to our web site for specific examples of two invasive species and their impacts: the zebra mussels (Figures 2 & 3), which are spreading beyond their original infestation sites in the Great Lakes, and the veined rapa whelk, which has been spotted in the lower reaches of Chesapeake Bay.

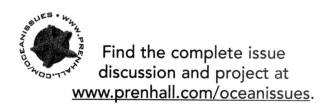

Find the complete issue discussion and project at www.prenhall.com/oceanissues.

[1] The U.S. Bureau of Reclamation operates a series of giant pumps at Tracy California on the delta of the Sacramento River to ship Bay water to Southern California. During 1998, they found so many mitten crabs clogging fish screens (which keep fish out of the pumps and so keep them from being cut to pieces) that the agency spent $400,000 to build a series of "crab screens" that catch the crabs before they clog the fish screens. The devices then fling them onto a conveyor belt for removal by a firm that pays for the privilege. The firm uses the mitten crabs for bait. Ironically, Chinese mitten crabs are a treasured delicacy in Hong Kong, and some entrepreneurs have explored shipping the California crabs to Hong Kong. But the state refuses to allow this, fearing that it will encourage more importation of the crabs and other exotic species, with more unforeseen consequences! (San Jose Mercury News. Crab Migration Drops Off, by N. Vogel. Oct 14, 1999.)

[2] www.wsg.washington.edu/pubs/bioinvasions/bioinvasions5.html

[3] www.darwin.bio.uci.edu/~sustain/bio65/lec09/b65lec09.htm

[4] www.ucc.uconn.edu/~wwwsgo/ballast.html

[5] www.serc.si.edu/invasions/ballast.htm#Box 1

Figure 2

The zebra mussel, shown here covering a crayfish. This invader has spread to 20 U.S. states and can tolerate estuarine water. (*GLSGN Exotic Species Library*)

Figure 3

Zebra mussels clog a water intake pipe. Industry and government over $70 million for research, control, and management of this species. (*Craig Czarnecki* [top], *Don Schloesser* [bottom]).

Sharks

- Would it matter if sharks were to disappear from the ocean?
- Is this an event that could happen?
- How do scientists determine the status of shark populations?
- What roles do sharks play in their environment?
- What are policy makers doing to protect sharks?

Background

There is one feature of the biology of sharks that has become legendary—their dominance as oceanic predators. Four hundred million years of evolution have endowed sharks with a suite of adaptations ranging from an array of acute senses to a jawfull of teeth that are replaced before they can dull. Like the big cats of Africa and Asia, sharks are supreme hunters. But evolutionary adaptations do not guarantee success. Sharks, perceived by most swimmers as their nightmarish nemesis at the beach, ironically have themselves become threatened with extinction (Figure 1).

By 1998, the number of sharks along the East Coast of the U.S. had declined so precipitously that, in April of that year, the National Marine Fisheries Service (NMFS), in accordance with the Magnuson-Stevens Fishery Conservation and Management Act[1], implemented a revised Fishery Management Plan for 39 species of sharks (and other highly migratory fishes like swordfish and tuna). One year later, the plan was amended[2] and the number of shark species protected was increased. These plans were designed to replenish stocks of sharks that had become severely depleted.

Other countries and international organizations are also taking actions to protect sharks. The Department of Fisheries and Oceans of Canada issued an Atlantic Shark Management Plan in 1997. In 1999, member countries of the Food and Agricultural Organization of the United Nations endorsed an international shark management agreement.

Most recently (October 27, 1999), the U.S. House Committee on Resources unanimously approved a resolution opposing the wasteful practice of shark finning after it came to light that 55,000 blue sharks were killed in 1998 for their fins, which are the key ingredient in shark-fin soup, an Asian delicacy. Days later the entire U.S. House of Representatives also unanimously endorsed the ban.

Why has there been so much concern for sharks? What events are responsible for the plight of sharks? What does "severe depletion" mean and can it lead to extinction? Would the consequences of shark extinctions be entirely negative, especially since they consume economically important fish and they also periodically "terrorize" swimmers? What roles do sharks play in their environment? In this issue we will address these questions and discuss shark ecology and conservation.

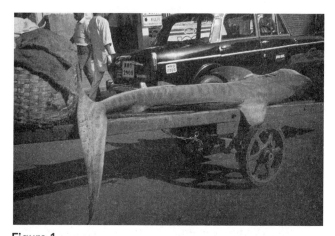

Figure 1

Tiger shark on a cart in Bombay India. Now a protected species along the U.S. Atlantic and Gulf coasts, this species is still threatened due to overfishing. *(BarnabasBosshart/CORBIS)*

Characteristics of Sharks

Sharks are fish. Thus, like tuna, mackerel, salmon, etc., they are members of the Phylum Chordata[3] and Subphylum Vertebrata[4]. However, sharks and their relatives (skates and rays) are collectively known as elasmobranchs[5]. They differ from tuna, mackerel, and salmon, which are bony fishes, while sharks, skates, and rays are cartilaginous fishes. These terms distinguish the principal material comprising the skeleton[6].

The following list compares some of the characteristics of sharks (Figure 2) with bony fish.

- Sharks have 5–7 external bilateral gill slits, while bony fish have a single bilateral opening covered by bony opercula.
- Sharks may have spiracles (small openings on the top of the head leading to the gill chamber).
- Sharks have ventral mouths, that is, underslung jaws, with some exceptions, most notably the recently discovered shark known as megamouth. The mouths of bony fish can be at the front (terminal), ventral (facing downward) or dorsal (facing upward).

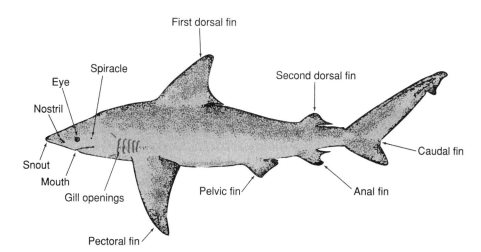

First dorsal fin

Spiracle

Eye

Nostril

Snout

Mouth

Gill openings

Pectoral fin

Second dorsal fin

Caudal fin

Pelvic fin

Anal fin

Figure 2

Shark external anatomy.

- Sharks have a tail fin whose upper lobe extends farther up and out than the lower lobe. Bony fish have a variety of symmetrical and asymmetrical tail shapes.
- Sharks have sandpaper-like skin due to the presence of placoid scales, which are structures resembling small teeth. Scales on bony fish are small bony plates.
- Sharks employ internal fertilization via claspers, the primary male sexual accessory organ. Very few bony fish use internal fertilization. Instead they release sperm and eggs into the water, and fertilization is external.
- Sharks lack a swimbladder, an organ that regulates buoyancy.
- Sharks reduce the stress of living in a salty medium by conserving and storing urea, which is a waste product eliminated by other organisms.

Sharks and sharklike ancestors have been on the planet for nearly 400 million years, and there are over 400 extant[7] species. Contrary to popular opinion, they are not "living fossils." While several features of sharks, especially the shark look, have been retained from their early evolutionary origin, many other characteristics, such as advanced jaw suspension, which enables them to feed more efficiently, and more flexible fin structure, which allows more maneuverability, have arisen more recently.

Ecology of Sharks

In coastal and oceanic ecosystems, larger sharks occupy a position at the top of the food web. Thus, they are apex, or top, predators. Within an ecosystem, apex predators exert a strong influence on the organisms below them in the food web. First, they play a major role in controlling both the diversity and abundance of other species within a community. Second, they influence the evolution of other species. In both cases they do this through predation on sick, injured, slow, weak, or otherwise less fit individuals.

Threats to the Survival of Sharks

For any species or population of organisms there exists a population size below which that group is believed to be doomed to extinction. This is known as a population's or species' **critical number**. When the number of adult individuals drops to this critical level, there is simply not a large enough reservoir of genetic variability and potential for mating to allow the population or species to propagate and successfully face the rigors of its environment, including competition with other species. Genetic variation is central to the survival of a species because it is the raw material for natural selection. When a population's environment changes, genetic variation may produce some individuals which have the characteristics necessary for survival.

Critical numbers are difficult to establish and efforts to do so may not be undertaken until a species is near extinction. They also vary among species and are influenced by life history characteristics. The California condor, the big cats of Africa, India, and Asia, many whales, and the desert pupfish are examples of organisms that may be at or below their critical numbers.

Many species of sharks may be threatened with extinction because of overfishing and habitat alteration. Annual catches of sharks, skates and rays reached 800,000 metric tons (nearly 2 billion pounds) by 2000[8]. Fishing pressure is likely to increase because most shark fisheries are still small-scale. As a transition to more modern, industrial-type fisheries is made, landings could increase substantially, at least for the short term.

Sharks are also frequently captured on longlines and in nets as bycatch (Figure 3), that is, as untargeted and hence unused catch. For example, in the 1991 Japanese squid driftnet fishery, which has subsequently ceased (although driftnets are still used in other fisheries), at least 11 kinds of sharks were taken as bycatch, including nearly 100,000 blue sharks[9].

Sharks are particularly vulnerable to pressures that can reduce their population because of their life history characteristics:

Figure 3

Oceanic white-tip shark caught in a Japanese driftnet in the Tasman Sea. *(Greenpeace/Grace)*

- There are typically fewer individuals of sharks than other species in the community.
- Sharks are very slow–growing.
- Individuals are long–lived and reach sexual maturity late in life.
- There is a long gestation period[10].
- Sharks typically rely on specific mating and nursery areas.
- Fecundity[11] is low.

Case Study 1: The Sandbar Shark

Consider the sandbar shark, *Carcharhinus plumbeus*, a large coastal species found worldwide and at one time abundant from Cape Cod to Brazil in the western Atlantic Ocean.

Before fishery scientists can determine if a species is threatened with extinction, it is critical to understand its life history. For some sharks, including the sandbar shark, these are at least partially known. Estimates of age and growth parameters for sandbar sharks have been obtained from tagging studies and research on captive animals. These results show that sandbar sharks are born at a length[12] of about 45–50 cm (16–20 in) and grow about 8–10 cm (3– 4 in) the first year. They reach sexual maturity at about 136 cm (53.5 in) at 15 years of age and have a life span of about 30 years. Eight to ten pups are born every other or every third year after a 12-month gestation period. In contrast, a typical bony fish such as a trout or cod reaches maturity earlier, has more young, grows faster, and has a considerably larger population size.

The most complete data on population levels of sandbar sharks, the most common shark in Chesapeake Bay, are from the long-term program of the Virginia Institute of Marine Science (VIMS) shark ecology program. This program sampled 8 or more stations in and offshore of Chesapeake Bay from May through September or October annually over the period 1980 to present. Sharks were captured with 100-hook longlines[13] stretched for 1–1.5 nautical miles (1.8–2.8 km) and fished 3–4 hours[14]. We will analyze their data on our web site, where you should now go.

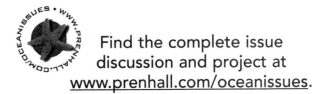

Find the complete issue discussion and project at **www.prenhall.com/oceanissues**.

---

[1] This can be viewed or downloaded from: http://www.wh.whoi.edu/magact/index.html

[2] This document, and updates, can be viewed or downloaded from: http://www.nmfs.gov/ Follow the links for "Atlantic Highly Migratory Species Management Plans".

[3] Members of the Phylum Chordata possess a notochord, dorsal hollow nerve cord, and pharyngeal gill slits at some point in their life history.

[4] The most prominent feature of vertebrates is the evolution of the primitive notochord into a skull and vertebral column (backbone).

[5] Pronounced "ee-laz-mo-branks"

[6] Bone is heavy, brittle tissue that is impregnated with hardening calcium salts. Cartilage, on the other hand, lacks these salts that contribute to bone's hardness, is fairly flexible and light, and does not contain the nerves and blood vessels that support the cells of bone.

[7] "Extant" is the opposite of "extinct" and refers to species still existing.

[8] Bonfil, R. 1994. Overview of world elasmobranch fisheries. FAO Fisheries Technical Paper 341, 119 pp.

[9] McKinnell, S. and M.P. Seki. 1998. Shark bycatch in the Japanese high seas squid driftnet fishery in the North Pacific Ocean. Fish. Res. 39: 127-138.

[10] "Gestation period" refers to the time developing embryos are retained in the female or in the egg.

[11] "Fecundity" refers to the number of offspring produced by a female of a species.

[12] Precaudal length, that is, from the tip of the head to the base of the tail. All data are from Dean Grubbs of the Virginia Institute of Marine Science (personal communication)

[13] Longlines are horizontally stretched ropes or heavy monfilament (the "mainline") with branches (the "gangions") containing hooks emanating from the mainline at fixed intervals.

[14] Although this method sounds dangerous, it generally does not harm the sharks, which are simply tagged and released. Commercial longlines, as you will later see, are a different matter.

Lifestyles of the Large and Blubbery: How to Grow a Blue Whale[1]

- What enables the blue whale to grow so large?
- Why is the Antarctic marine ecosystem so productive?
- What are krill and why are they considered a "keystone species"?
- What are the threats to the Antarctic marine ecosystem?

Introduction

What is the largest living organism on Earth?

If you answered "blue whale" you'd be wrong. The designation actually belongs either to the redwood tree *Sequoia sempervirens* or to the soil fungus *Armillaria bulbosa*, which lives in northern Michigan.

But the blue whale (*Balaenoptera musculus*; Figure 1) is the largest animal that ever lived, reaching a length of about 30 m (98 ft) and weighing around 100 tons (91,000 kg).

Have you ever wondered why this is the only living animal that grows to this size? Exactly what permits this species to achieve such dimensions but prohibits others from doing so? In this issue you will answer the last question yourself.

An Introduction to Antarctic Marine Ecology

The blue whale spends its summers in the waters surrounding Antarctica, the planet's fifth largest continent. Antarctica is completely surrounded by the southernmost extensions of the Pacific, Atlantic, and Indian Oceans, a physically distinct body of water collectively referred to as the Southern Ocean. The closest continent, South America is 970 km (600 mi) away.

Antarctica is, of course, a land mass, but you wouldn't know it from looking at it: nearly 98% of its 14 million km^2 (5.4 million mi^2) is covered with ice up to 4.5 km (nearly 3 mi) thick. Including sea ice, Antarctica holds 90% of the world's ice and 70% of the world's fresh water. Ironically, despite this much fresh water, because there is no rain in the interior (less than 4 cm of fine crystals known as 'diamond dust' per year!), the continent is essentially a barren desert.

The Southern Ocean

In contrast to the Antarctic land mass, the Southern Ocean supports a richly productive assemblage of organisms.

What makes these cold waters so productive is the southerly flow of deep, nutrient-rich water known as Circumpolar Deep Water[2]. This water rises to the surface (upwells) near the continent from depths of 3000 m (9800 ft) and makes nutrients available to photosynthetic or-

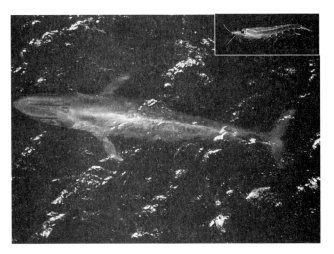

Figure 1

Blue whale, the largest animal that ever lived, and a major consumer of Antarctic krill. *(Courtesy of M. Carwardine/Still Pictures)* **Figure 2** (Inset) Antarctic krill, a keystone species in the Southern Ocean. *(Photo by Frank T. Awbrey/Visuals Unlimited)*

ganisms (principally diatoms), which must live in the sunlit surface waters. It also transfers heat to the Antarctic atmosphere. In addition, during the Antarctic summer there is six months of daylight. Thus, photosynthesis can occur 24 hours a day during the summer.

The northern limit of this productivity is a narrow belt of water from 20 to 30 miles (32–48 km) wide known as the Antarctic Convergence, or Polar Front. Here, cold Antarctic surface waters sink below warmer waters flowing south. This is also the northern boundary of the Southern Ocean, and it is what makes the Southern Ocean 'physically distinct'.

Krill

The key indicator of this Antarctic productivity, in terms of biomass, is not the diatoms or other tiny photosynthetic organisms that actually use the nutrients in the first place. It is a 5 to 6 cm (2–2.5 in) long crustacean called Antarctic krill (*Euphausia superba*, Figure 2) which can top the scales at 2 g (0.1 ou). Krill, meaning "young fish" in Norwegian, is a term applied to a group of about 85 species of shrimplike organisms collectively called "*euphausiids*"

(Phylum Arthropoda) that inhabit waters from the poles to the tropics. Krill are filter-feeding organisms which use their setae (bristle) covered appendages to trap the tiny diatoms. They are heavier than water and must swim continuously to avoid sinking. While the life span of krill is not known with certainty, scientific estimates range from 5–11 years. Krill are eaten by whales, seals, birds, squid, and fish, and to a lesser but growing extent by humans.

It is quite possible that krill are the most abundant animal species on earth[3]. Dense aggregates weighing an estimated 2 million tons (1.8 billion kg) and covering as much as 450 km^2 (170 mi^2) have been observed. Density of krill schools reportedly reach 30,000 animals per cubic meter of seawater[4]! The total standing stock (the biomass at any time) has been estimated to be between 200 and 700 million tons (200–700 billion kg). In contrast, the 1997 total world fish catch was less than 100 million tons (100 billion kg).

Thus, the numerical abundance and biomass of krill, as well as its important position in the food web (see below), make it a *keystone species* in its ecosystem. A keystone species is one whose impact on its ecosystem is disproportionately large and whose loss would severely disrupt the system.

Krill Fisheries

Before the advent of European whaling in the 19th century, baleen (filter-feeding) whales consumed huge volumes of krill. After the near-extinction of these whales by the 1960s, some viewed krill to be present in "surplus" amounts (estimated at 150 million tons!) which could be harvested without impacting either the species of the marine environment.

Activity: Comment on this idea of "surplus food" in an ecosystem and the impact of harvesting it.

Large-scale fisheries for krill (Antarctic and North Pacific krill) currently occur only in Antarctic waters and off the coast of Japan. In the mid-70s the Soviets were the first to operate full-scale fisheries for Antarctic krill. Japan, Poland, and Ukraine currently conduct Antarctic krill fisheries.

As conventional world fisheries decline from overfishing, and as aquaculture continues to grow, there will be greater emphasis on increasing commercial catches of species from more distant waters. This increase in demand will lead to a greater pressure on Antarctic krill, which are the largest known krill stock.

Currently, most krill taken are used as bait and aquaculture feed, although a significant proportion is used for human consumption. Krill are fed to farmed fish because of their nutritional value and also because krill contain high concentrations of the red pigment group carotenoids, which heightens the red color of fish. Japanese consumers consider red as a sign of good luck and as an appetite stimulant.

The Antarctic krill fishery is not increasing currently due to the immense expense of conducting a fishery in Antarctic waters. However, it is likely that in the future this will change because of the growing pressure to catch krill commercially for use as feed in aquaculture. Krill fisheries may also respond to demand for krill concentrate, a health food because of its omega-3 fatty acids content. What this portends for krill is not known because fishery managers have little experience in dealing with an organism such as the krill, which occupies a low but vital position on the food web[5]. Thus, while krill fisheries may increase the world's ocean harvest, they also could cause harm, perhaps catastrophic harm, to this species and to marine ecosystems.

Energy Transfer in the Ecosystem

As mentioned above, the cold Antarctic waters are extremely productive. This productivity is fueled by the upwelling of nutrient-rich water and the long photoperiod. It also depends upon the number of links in the individual food chains comprising its food web. Biological processes are not 100% efficient. Thus, at each step in a food chain, when one organism consumes another, only a small fraction, typically 10 to 20% of the energy value of the food, winds up being incorporated into the body of the grazer or predator. The rest of the energy is "lost" into the environment, primarily through the process of respiration. Thus, the more steps there are between the photosynthetic organisms at the base of the food web and the top predator, the greater the energy loss.

Figure 3 depicts a simplified part (the pelagic portion) of an Antarctic food web. One pathway leads from diatoms to krill to squid to Weddell seals to leopard seals to orcas (a toothed whale). This pathway thus has six levels with five intervening steps. A second pathway has three levels and two steps: diatoms to krill to blue whales (which are filter feeders).

This figure shows that krill are eaten by many of the winged birds, penguins, fishes, squid, and by one seal. Two species of penguin, the Adelie (which is the most abundant penguin in Antarctica) and the chinstrap, live predominantly on squid. The population sizes of these two species are orders of magnitude greater than the larger emperor and king penguins, whose diet is more diverse.

A similar situation occurs in seals. The most common seal, the crabeater, has a diet based almost exclusively on krill. Its population size, estimated at 15 to 40 million, is greater than that of all the other Antarctic seals combined. The blue whale illustrated in this food web is not the only krill-feeding whale in Antarctic waters. Right, minke, sei, fin, and humpback whales are also found in the Southern Ocean.

We'll compare these two pathways on our web site.

Threats to the Antarctic Marine Environment

Even though it is far removed from the kinds of environmental problems most of the world is facing (fisheries notwithstanding), the Southern Ocean ecosystem is facing

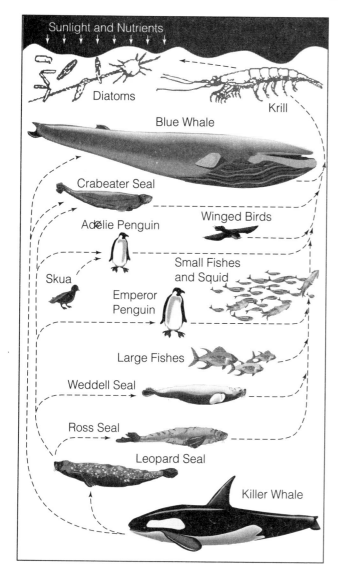

Figure 3

Simplified Antarctic marine food web.

Figure 4

Icebergs being towed from shipping lanes in Antarctica. Will this costly procedure be more commonplace as temperatures increase due to global warming?   *(Photo by Steve Berkowitz)*

Antarctic organisms, at least in the short term. Around 1985, scientists became aware that ozone levels in the Antarctic stratosphere were dropping significantly each spring. The source of the degradation was a group of synthetic chemicals called chlorofluorocarbons (CFCs), which were, and still are, used as propellants and refrigerants. Under the Antarctic atmospheric conditions (extreme cold and constant sunshine), these chemicals acted as powerful ozone depleters. The result of this ozone depletion has been an increase in ultraviolet radiation reaching the surface of the Earth and penetrating the top meter or so of seawater. High intensities of ultraviolet radiation have the potential to decrease phytoplankton production by an order of magnitude or more.

Now, go to our web site to complete this analysis.

some major threats. Temperature increases due to global warming will likely result in accelerated melting of ice sheets (Figure 4) and the addition of massive volumes of freshwater to the marine environment. This could disrupt the sinking of cold, dense water masses (known as thermohaline circulation), which is a major source of the world's bottom water. The effects of disrupting the flow of bottom water cannot be predicted, but they would certainly be severe. Also, Antarctic marine organisms would be particularly susceptible to temperature increases because they live (and have evolved in) a relatively stable thermal environment. In addition, like all organisms, Antarctic organisms live closer to their thermal maximum (the highest temperature they can tolerate) than their thermal minimum. Thus, increases in temperature could lead to a host of lethal and sublethal effects.

As bad as the effects of global warming might be, the thinning ozone layer may be a more pressing problem for

Find the complete issue discussion and project at <u>www.prenhall.com/oceanissues</u>.

---

[1] Written with Steve Berkowitz of the Department of Marine Science of Coastal Carolina University

[2] In the Atlantic this is known as "North Atlantic Deep Water" and is similarly named for the Indian and Pacific bodies.

[3] A small deepwater fish, *Cyclothone*, may rival krill in abundance.

[4] Nicol, S. and Y. Endo. 1997. Krill fisheries of the world. FAO Fisheries Technical Paper 367. 100 p.

[5] ibid

Beaches or Bedrooms? The Dynamic Coastal Environment

- What factors affect beach location and shape?
- How has coastal development influenced the beach environment?
- What are the pros and cons of beachfront stabilization?

Background

Beaches are dynamic zones that change in shape over both space and time. Beaches may stretch for thousands of km along passive margin shorelines (i.e., those without seismic activity such as the U.S. Atlantic coast), but extensive beaches are rare in tectonically active uplift areas along active margin shorelines, for example, the Pacific coast of North and South America.

The shape of a beach may vary over distances of a kilometer to hundreds of kilometers and is based on the balance between processes that promote erosion and processes that favor deposition of sediment. These in turn are influenced by local physical and geologic characteristics.

Over time, the shape of beaches is influenced by short-term storm events (days); normal seasonal changes in wave intensity (month to year); longer-term climatic phenomena like El Niño and La Niña (years to decades) and periods of high hurricane activity; and very long-term (century to millennium) changes in global sea level.

In the last 18,000 years, sea level has risen about 120 meters (400 ft) and many coastal shorelines have moved dramatically inland. Off the East Coast of the United States geologists estimate that the shoreline has migrated inland more than 80 km (50 mi) during this period, alternating between active migration and static periods. Based upon a rough average of typical beach slopes, for each unit rise in sea level, the beach will migrate 1000 to 2000 units inland.

All 30 states bordering an ocean or the Great Lakes have erosion problems, and 26 are presently experiencing net loss of their shores. As an example of this erosion and shoreline migration process, consider the Cape Hatteras lighthouse in North Carolina and its location relative to the shoreline. In 1870, the lighthouse was situated about 450 m (1475 ft) from the ocean. By 1919, the ocean had advanced to within 100 m (325 ft) of the tower, and by 1935, it was within 30 m (100 ft). After a series of strategies designed to hold back the coastal erosion failed (not a surprise to coastal geologists), the lighthouse was finally moved nearly 900 m (2950 ft) inshore in 1999. Attempts to save the neighboring Morris Island lighthouse in South Carolina are currently in progress.

Erosion is a natural and sometimes predictable process. Despite this knowledge, the movement of shorelines inland has been a serious problem to coastal residents and businesses, especially in the middle and southern Atlantic and Gulf regions of the United States. This issue is further examined on our website.

Composition and Structure of Beaches

Many factors affect the shape, composition, and structure of beaches. The shape varies with sand supply, sea level change, and wave size. When any of these factors change, the other factors respond accordingly. For example, if the supply of sand is reduced and sea level and wave size remain constant, the beach profile will change because the diminished sand supply creates an erosional environment, leading to the landward movement of the beach's intertidal region. Decreasing wave energy, holding the other physical factors constant, will cause either reduced erosion or natural accretion (building) of the beach.

The slow but unrelenting assault of sea level rise upon coastlines is difficult to recognize on a daily basis, but storm or hurricane-induced erosion, property damage, and loss of life capture our immediate attention. Fortunately, increased capabilities in weather forecasting have resulted in a decrease in the loss of life from storms over the last century in the United States. However, increasing population growth and development in coastal regions has led to an increase in property damage from extreme weather events over this same time period (Figure 1; http://hurricanes.noaa.gov/prepare/).

It has been predicted that global sea level will rise by approximately 1 meter in the next century, meaning that along the U.S. Atlantic and Gulf coasts, the shoreline will migrate up to 2000 meters (6560 ft) inland. Compound this phenomenon with the anticipated increased hurricane activity in the Atlantic, and the long-term prognosis for coastal development begins to look precarious.

To further complicate matters, human attempts to slow beach erosion often exacerbate the problem. Construction of seawalls, jetties, revetments, and groins may stabilize local portions of the beach but can cause problems in adjoining areas. Seawalls, which parallel the shoreline, may temporarily protect property but eventually lead to the loss of the beach, a process called "Newjerseyization," named for the state where seawall construction has been a common practice in postponing property loss (Figure 2).

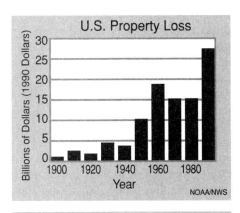

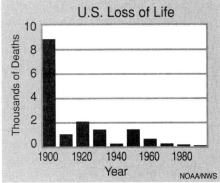

Figure 1

The graphs above from the NOAA National Weather Service illustrate the great increase in property damage but reduction in the loss of life associated with U.S. hurricanes since 1900. *(NOAA)*

Figure 2

Newjerseyization is evident at high tide in North Myrtle Beach, South Carolina as erosion has removed the beach up to the seawall. Inset, erosion has excavated the foundation to a modern beachfront high rise in North Myrtle Beach, South Carolina. *(Photos by Abel)*

Beach Renourishment

Although many states no longer permit construction of hardened structures such as sea walls and groins, the alternative practice of beach "renourishment" has become increasingly common in the last several decades. Beach renourishment involves bringing in sand, by dredge, barge, or truck, to build up the height and breadth of disappearing beaches. The sand may come from inland sources, or from subtidal shelf regions. This practice is expensive and temporary, but it provides a key benefit compared to constructing hard structures in that it allows the beach to continue to exist. This benefit of beach renourishment, combined with a significant federal subsidy, is an especially appealing alternative in areas which depend on tourism.

The U.S. Army Corps of Engineers is responsible for designing and implementing renourishment operations. Much controversy exists regarding the efficacy and accuracy of the Corps' renourishment practices. One of the most common problems is that the Corps overestimates how long a beach will last after renourishment. As of 1987, 400 million cubic yards (365×10^6 m³) of sand had been used to renourish beaches in the U.S. By 1995, the total cumulative national cost of continued renourishment approached $5 billion.

Renourishment practices have led to a false sense of security for many beachfront residents, and the continued subsidy of renourishment by the federal government has had the effect of stimulating additional beachfront development. Technically, we may be able to continue in the near future to repair the localized storm impact erosion to prime tourism locales (win the battle), but the reality of global sea level rise means that in the future we will eventually have to abandon current beachfront property (lose the war), or install an extremely expensive series of dikes, as they have done in Holland, to protect our shoreline. The Dutch have spent around $15 billion during the 90s, which is around $1,000 for each citizen, suggesting that we might have to spend $270 billion to "protect" our coasts, a sum roughly equal to one year's federal military budget.

Impacts on Fauna

Beach renourishment may have both direct and indirect effects upon flora and fauna of the sandy shore. The direct effects of beach renourishment include impacts that occur during dredging of subtidal materials, as well as those that occur during emplacement. Many coastal dredging operations which obtain sand material for beach renourishment impact species in the process. Certain types of dredges can cause high mortalities of female sea turtles during the nesting season. Many of the subtidal invertebrate organisms in the dredge material die after emplacement into the intertidal environment.

Among the impacts which occur after emplacement are changes in the food webs of the beach sands, as well as influences upon temperature and humidity of sand in the high intertidal region of the beach where sea turtles often lay their nests. Cinde Donoghuc, a graduate from the University of Virginia Environmental Science program, has recently studied renourishment effects on Pea Island (near Nags Head, North Carolina). She discovered that renourishment caused an immediate decrease of at least 40 percent in mole crabs, coquina clams, and ghost crabs. This reduction lasted between 2 to 5 months. However, even longer recovery (up to 8 months) was required if the dredged material was placed on the beach during peak larval recruitment times. Many shorebirds are dependent upon these common beach invertebrates for food.

The concern that exists for sea turtles is based on the impacts that altering the grain size, organic matter content, and sediment color have upon the processes that take place during egg incubation. Sea turtles have been shown to select certain regions of the beach to lay their eggs. Critical to nesting is their ability to excavate the sand to produce an egg chamber into which the eggs are delivered. The success of hatching of eggs is dependent upon a narrow range of temperature and humidity conditions within the nesting sand. Changes in grain size and color influence these factors and may deter nesting or reduce hatching success.

Renourishing beaches with inland sand sources often results in mismatched grain size, color, and organic matter content. In addition, bacterial and fungal spores from terrestrial sand sources may have more pathogenic components than subtidally-obtained renourishment sources.

The process of renourishment probably has even more severe indirect effects, namely promoting an attitude that beachfront development can be pursued anywhere and everywhere. A variety of species of animal and plants that used to inhabit the coasts are now endangered or threatened because of elimination of their habitat. For example the threatened loggerhead sea turtle (*Caretta caretta*) has experienced a dramatic population decline in the southeastern states of South Carolina and Georgia in part due to beachfront development. Many species of shore birds like the Piping plover (*Charadrius melodus*) are sensitive to dune development. The dramatic decline in this species (a loss of 1,240 breeding pairs along U.S. Atlantic coast as of 1990) caused it to be added to the federal list of "Threatened and Endangered Species" in 1985, and it is protected under the "Endangered Species Act." Go to our website to continue the analysis of this issue.

Find the complete issue discussion and project at www.prenhall.com/oceanissues.

Greenhouse Gases, Global CO$_2$ Emissions, and Global Warming

- What is the composition of the earth's atmosphere?
- What processes have influenced the atmosphere's composition?
- How does the earth's atmosphere interact with the ocean?
- What are greenhouse gases?
- What will be the impacts of global climate change associated with greenhouse gas increases?

Introduction

Earth's atmosphere is a relatively thin shell. About 95% of it is contained within 14 km (8.6 mi) of the earth's surface. All life on earth depends on the atmosphere, but despite this knowledge, humans are altering the atmosphere's composition. Scientific evidence suggests that emissions from the burning of fossil fuels, from industrial sources such as cement manufacture, and from deforestation are changing the make-up of our atmosphere. In addition, trace gases such as methane and chlorofluorocarbons (CFCs and HCFCs) are having an impact on the atmosphere wholly out of proportion to their concentration.

The Earth's Early Atmospheres

Our current atmosphere is thought to be the earth's third over its cosmic history. The first was comprised of light gases, primarily hydrogen (H$_2$) and helium (He$_2$). Slightly heavier inert gases such as argon (Ar$_2$), krypton (Kr$_2$), and neon (Ne$_2$) were also likely present, along with dust. This first atmosphere was blown away as the sun reached critical mass and began its internal fusion reactions, and by heat generated by the early molten earth between 4 and 5 billion years ago.

The second atmosphere was formed between 3.8 and 4 billion years ago when gases escaped from the earth as its crust solidified. It is also widely believed that bombardment by comets contributed gases to the earth's second atmosphere. This second atmosphere contained a variety of gases but was devoid of oxygen, a gas necessary to support most of the earth's present life forms.

The Development of the Current Atmosphere

The earth's third and current atmosphere was produced partly by the metabolism of living organisms. Photosynthetic organisms, the cyanobacteria, appeared about 3.8 billion years ago and began to produce oxygen. Over the next 2 billion years, atmospheric O$_2$ concentrations rose and CO$_2$ concentrations fell as a direct result of photo-synthesis along with the accumulation of carbonate rocks. This increase in atmospheric oxygen then set the stage for two major evolutionary events on the planet: the evolution of aerobic (oxygen-using) life forms and the establishment of the ozone layer.

Increases in atmospheric and seawater oxygen levels poisoned sensitive microorganisms and led to the evolution of other microorganisms capable of using oxygen to liberate energy from organic compounds. This new *aerobic* metabolism was much more efficient than the *anaerobic* metabolism it replaced. Today, natural anaerobic microbial communities are restricted in their habitats to deep marginal or isolated seas, organic-rich sediments, and regions beneath the earth's surface. The development of aerobic metabolism also permitted the evolution of multicellular organisms, which required more energy to support their increased biomass. All existing multicellular life forms employ aerobic metabolism.

Increases in atmospheric oxygen concentration eventually triggered the formation of the layer of ozone (O$_3$) that now exists in the upper atmosphere. Although considered a pollutant at ground-level, the ozone layer serves as an important filter of harmful ultraviolet radiation, which is responsible for sunburn and can cause skin cancer in humans.

Prior to the existence of the ozone layer, the earth's terrestrial and ocean surfaces were exposed to extremely high intensities of ultraviolet radiation. The surface ocean waters filtered out some of this radiation, and thus provided some protection to organisms, but it is likely that ocean primary production (that is, production of high-energy compounds from photosynthesis) was still limited by the high ultraviolet light intensities. The terrestrial surface fared worse and was likely sterilized by this radiation.

The Atmosphere's Current Composition

The present-day atmosphere is composed primarily of N$_2$ gas (78.08% by volume), and oxygen (20.94% by volume). Also present, in quantities less than 1% by volume, are, in order: Ar, H$_2$O, CO$_2$, Ne, He, CH$_4$, NO$_2$, CO, NH$_3$, and O$_3$. The major controls upon the composition of the atmosphere and the cycling of these compounds are interactions with the Earth's biosphere (living matter) and lithosphere (rock and geological processes such as volcanism). Presently, O_2 levels seem to be stable, but CO_2 levels are not and display seasonal and longer-term trends. Seasonal changes in CO_2 concentration are related to primary production changes due to changing light durations. Longer-term (decade–century) increases in CO_2 are due to a variety of anthropogenic (human-caused) inputs as well as changes in land use that reduce the ability of terrestrial biota to absorb CO_2.

Because the amount of CO_2 in the atmosphere is relatively small, the concentration is easily changed by the addition of CO_2 from various sources.

Atmospheric Functions

In addition to providing the oxygen needed by most of earth's life forms, the atmosphere provides a significant thermal insulation, preventing extreme changes in temperature over the daily light-dark cycle. Unequal heating of the earth's atmosphere and terrestrial surface create long-term climate and short-term weather patterns. The winds that result from these heating differences and resultant pressure differences also drive ocean currents. The atmosphere also transfers heat from the tropics toward the poles.

Changes Caused by Humans

After decades of research, scientists have finally concluded that humans have changed the composition of the earth's atmosphere. Since the beginning of the industrial revolution, atmospheric concentrations of the following greenhouse gases have changed: carbon dioxide has increased 30%; methane concentrations have more than doubled; and nitrous oxide concentrations have risen by about 15% (Figure 1). Greenhouse gases allow short wavelength radiation (mainly visible light) from the sun to pass through the atmosphere, but they absorb the longer wavelength radiation (i.e., heat or infrared) that is emitted by the earth. The need for energy to support industrial development, heat homes, cook food, watch television, and surf the in-

ternet, as well as the increased use of automobiles, has resulted in the burning of great stores of fossil fuel.

Fossil fuels, including coal, oil, and natural gas, were formed by the preservation and slow anaerobic decomposition of ancient plant and phytoplankton deposits. These deposits took tens of millions of years to form but we are now utilizing them at a rapid rate. A key by-product of fossil fuel consumption is CO_2. Since the industrial revolution, we have added CO_2 to the atmosphere more rapidly than it can be absorbed by its variety of sinks. This has led to a slow but steady increase in CO_2 concentration that will result in at least a doubling of pre-1860 atmospheric CO_2 content by the year 2150, if present trends continue. The increase in atmospheric CO_2 has occurred not only because of increased inputs into the atmosphere, but also due to changes in the landscape, such as deforestation, that result in less removal of CO_2 from the atmosphere.

Methane is another greenhouse gas whose concentration has also become elevated because of human activities. It is emitted by cows, flooded farmlands (i.e., rice paddies), and landfills and all of these have increased dramatically within the last 2 centuries.

Global Warming Impacts

There is little controversy that human activities have caused these changes in the atmosphere, but there is intense debate regarding the magnitude of the problem that may result from these changes. However, knowledge that the 10 warmest years in this century all have occurred in the last 15 years has convinced most scientists that we are now seeing the direct response of the planet to the increase in greenhouse gases.

What will the impacts be of global climate change associated with greenhouse gas increases? Examine Figure 2 for an overview.

Human health will be directly affected by increases in the heat index, which can impact healthy people, but especially affects the elderly and those with heart and respiratory illnesses. Higher temperatures will also increase ozone pollution in the lower atmosphere, which also is a threat to people with respiratory illnesses. Global warming may also increase the incidence of some infectious diseases, particularly those that usually appear only in warm areas. Diseases that are spread by mosquitoes and other insects, including malaria, dengue fever, yellow fever, and encephalitis, could become more prevalent if warmer temperatures and wetter climates enable those insects to become established farther north.

Another important impact of global climate change will be alterations in precipitation patterns. Predicted changes in climate are expected to enhance both evaporation and precipitation in most areas of the United States. The net balance of these processes influences the availability and quality of water resources. In areas expected to become more arid, like California, lower river flows and lower lake levels could impair navigation, reduce hydroelectric power generation, decrease water quality, and reduce the supplies

Figure 1

Changes in the global concentration of greenhouse gases since the pre-industrial period. *(IPCC, 1995)*

	CO_2	CH_4	N_2O
Pre-industrial concentration	280 ppmv	700 ppbv	275 ppbv
Concentration in 1994	358 ppmv	1720 ppbv	312 ppbv[2]
Rate of concentration change[1]	1.5 ppmv/yr	10 ppbv/yr	0.8 ppbv/yr
Atmospheric lifetime (years)	50-200[a]	12[b]	120

ppmv = part per million by volume; ppbv = part per billion by volume

[1] Concentration increases in CO_2, CH_4, and N_2O are averaged over the decade beginning in 1984.

[2] Estimated from 1992-1993 data.

[a] No single lifetime for CO_2 can be defined because of the different rates of uptake by different processes.

[b] Defined as an adjustment time which takes into account the indirect effects of methane on its own lifetime.

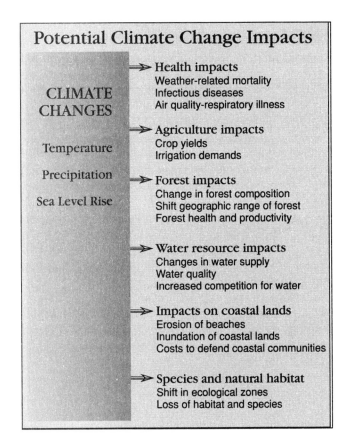

Potential Climate Change Impacts

CLIMATE CHANGES

Temperature

Precipitation

Sea Level Rise

➤ **Health impacts**
Weather-related mortality
Infectious diseases
Air quality-respiratory illness

➤ **Agriculture impacts**
Crop yields
Irrigation demands

➤ **Forest impacts**
Change in forest composition
Shift geographic range of forest
Forest health and productivity

➤ **Water resource impacts**
Changes in water supply
Water quality
Increased competition for water

➤ **Impacts on coastal lands**
Erosion of beaches
Inundation of coastal lands
Costs to defend coastal communities

➤ **Species and natural habitat**
Shift in ecological zones
Loss of habitat and species

Figure 2

Impacts of climate change. *(U.S. EPA)*

holes" found in the Northern Great Plains. A drier climate would decrease the amount of open water ponds in this region, with a commensurate reduction in duck populations. Among fish, those which inhabit inland aquatic environments are expected to be more vulnerable than coastal or marine species. Lake-locked fishes have little recourse in seeking cooler waters. Fish that inhabit rivers may be able to migrate northward to seek cooler water, but fish in east–west oriented rivers, like the Ohio and Danube will not be able to escape warming impacts. The results of a 1995 EPA study, "Ecological Impacts From Climate Change: An Economic Analysis of Freshwater Recreational Fishing," suggest that the overall diversity of fishes in U.S. rivers and streams is likely to decline, because of the loss of cold water forms.

Now, go to our web site to continue this analysis.

Find the complete issue discussion and project at www.prenhall.com/oceanissues.

of water available for agricultural, residential, and industrial uses. Lower water levels would also affect the biota. In other areas, increased precipitation, which is expected to be more concentrated in large storms as temperatures rise, will increase the incidence of flooding.

Shifts northward in climatic regimes would also affect the types of crops that can be raised. Studies conducted in the 1980s generally concluded that climate change would have severe impacts on agriculture. More recent assessments have suggested that these might be partially offset, at least in the U.S, by longer growing season and enhanced production from higher CO_2.

Climatic change will cause a shift in natural community composition as plant and animal species try to migrate to maintain their preferred habitats. For example, the projected 2°C (3.6°F) warming that will occur this next century could shift the ideal range for many North American forest species by about 300 km (200 mi.) to the north. Many plant species lack seed dispersion mechanisms that can adjust this rapidly, and thus these species may be susceptible to regional extinction. Coastal wetland plant communities may lose habitat because they may not be able to keep up with the predicted enhanced rate of sea level rise, or the transgression paths of these communities may be blocked by human development.

Among wildlife, certain birds and fishes are expected to be greatly impacted by predicted climate change. A large number of duck species are dependent upon "Prairie Pot-

Toxic Chemicals in Seawater

- What are the effects of polychlorinated biphenyls and tributyl tin on marine organisms?
- What are their concentrations in seawater and in the tissues of marine organisms?

Introduction

Polychlorinated biphenyls (PCBs) are a group of over 200 synthetic chemicals of a type called **organochlorines**. Their formula is complex and their atomic weight depends on the number of relatively heavy chlorine atoms in the structure. PCBs do not exist naturally on earth; they were originally synthesized during the late nineteenth century. Because of their stability when heated, they were widely used in electrical capacitors and transformers. In the 1960s scientists began to report toxic effects on organisms exposed to PCBs, and by 1977, the manufacture of PCBs was banned in the U.S., the U.K., and elsewhere. By 1992, 1.2 million metric tons (2.6 billion pounds) of PCBs were believed to exist worldwide, while 370,000 metric tons (810 million pounds) were estimated to have been dispersed into the environment[1].

PCBs are relatively heavy molecules (average atomic weight of around 360 g/mole) and are relatively insoluble in water. Concentrations in seawater may reach 1 part per million (ppm), but PCBs typically concentrate in sediment. From there, they enter the food chain mainly through the activities of organisms called sediment- or deposit-feeders, which eat sediment, extract organic matter, and excrete the rest. PCBs tend to accumulate in the fatty tissues of animals and thus can become concentrated in animal tissue, a process called *bioaccumulation*. If animals eat the deposit-feeders, the PCBs move up the food chain and become concentrated, a process called *biomagnification*. While seawater concentration is usually below 1 ppm, concentrations exceeding 800 ppm have been measured in the tissues of marine mammals. According to the Environmental Research Foundation, this would qualify the creature for hazardous waste status!

PCBs have become widespread and serious pollutants, and have contaminated most terrestrial and marine food chains. They are extremely resistant to breakdown and are known to be carcinogenic. PCBs have been linked to mass mortalities of striped dolphins (Figure 1) in the Mediterranean, to declines in orca (killer whale; Figure 2) populations in Puget Sound, and to declines of seal populations in the Baltic.

While PCBs may threaten the entire ocean, the northwest Atlantic is believed to be the largest PCB reservoir in the world because of the amount of PCBs produced in countries that border the north Atlantic.

Figure 1

Striped dolphins, whose mass mortalities in the Mediterranean Sea have been linked to PCBs. *(Photo by Erik Stoops, Corel Corporation)*

Figure 2

The killer whale, an organism at the top of the food web in which high levels of PCBs have been found, an example of bioconcentration and biomagnification. *(Photo by Francois Gohier)*

In terms of their specific effect on life, PCBs have been shown to cause liver cancer and harmful genetic mutations in animals. PCBs may inhibit cell division, and they have been implicated in reduction of plant growth and even mortality of plants.

According to a report edited by Paul Johnston and Isabel McCrea for Greenpeace UK,

"Since the rate at which organochlorines break down to harmless substances (has been) far outstripped by their rate of production, the load on the environment is growing each year. Organochlorines (including PCBs) are arguably the most damaging group of chemicals to which natural systems can be exposed."

PCBs and Orcas in Puget Sound

Even though soldiers during World War II used them for target practice, orcas have become a symbol of the Pacific Northwest.

In 1999, Dr. P.S. Ross[2], a research scientist with British Columbia's Institute of Ocean Sciences, took blubber samples from 47 live killer whales and found PCB concentrations from 46 ppm to over 250 ppm, up to 500 times greater than those found in humans. Ross concluded, "The levels are high enough to represent a tangible risk to these animals."

Ross compared the orca population he studied with the endangered beluga whale population of the St. Lawrence estuary of eastern North America, in which a high incidence of diseases have been linked to contaminants and which have shown evidence of reproductive impairment.

For the orcas, the PCBs are likely passed from generation to generation—an example of their persistence. PCBs are highly fat-soluble and are concentrated in mother's milk. Ross said, "…Calves are bathed in PCB-laden milk at a time when their organ systems are developing and they are at their most sensitive."

While PCBs have been banned in the U.S. for 20 years, they are still being used in many developing countries. Moreover, PCB-laden waste has been transported from industrial countries to developing countries for "disposal." (PCBs can be removed from soil or sediment by incineration, but it is expensive—too expensive for most developing nations). Approximately 15% of PCBs reside in developing countries, mostly as a result of shipments from industrialized countries. Accordingly, Ross speculates that PCBs in the Pacific could be derived from East Asian sources and could end up concentrated in the tissues of migratory salmon, which are a prime food source for the orcas.

Ross' study, one of the most comprehensive on cetaceans (whales and their relatives) to date, was done in collaboration with the University of British Columbia, the Vancouver, B.C. Aquarium, and the Pacific Biological Station of British Columbia.

J. Cummins, in a 1988 paper in *The Ecologist*, stated that adding 15% of the remaining stock of PCBs to the ocean could result in the extinction of marine mammals.

Tributyl Tin

Tributyl tin (TBT) is the name given to a number of compounds with the general formula $(CH_3(CH_2)_3)Sn$. Typical variations are TBT chloride and TBT cyanide. A typical molecular weight is 320 g/mole. Added to marine paint, the compound inhibits the growth of barnacles and algae on the bottom of boats. However, tributyl tin kills not only organisms in direct contact with the boat (at sub-part per billion levels) but is also lethal to other marine organisms as it seeps out of the paint. TBT also bioaccumulates in shellfish. These discoveries led certain nations such as Australia, France, the United States, and others to ban TBT starting around 1980.

However, TBT is still widely used on large ocean-going vessels, so trace amounts of it are found in almost every harbor. For the impact of tributyl tin on shipyard workers, go to http://www.ban.org/ban_news/shipbreaking_is.html. For a review of the nature and impact of tributyl tin, try http://www.mbhs.edu/class/teched/matsci/stephen.html.

TBT in Japanese Tuna

A university research team has found that tuna and bonito in the waters around Japan have concentrations of TBT many times that found in fish in the South Pacific and Indian Oceans, but not believed high enough to harm humans. This TBT is presumably from anti-fouling paints used on ship hulls as well as from material used to protect fish nets. The researchers concluded that, in addition to contaminating local fish, the seas around Japan are now sources of TBT contamination of migratory fish like bonitos and tunas.

Shinsuke Tanabe, a professor at Ehime University's Agriculture School, and his colleagues collected 47 tuna and bonito from the central Sea of Japan, the waters off Kochi Prefecture, Papua New Guinea, the Indian Ocean and five other areas from 1983 through 1996. Tuna caught in the central Sea of Japan were found to have the highest concentration of TBT–320 nanograms (1 nanogram is equal to 1 billionth of a gram) as well as two other forms of tins. Tuna caught off Kochi Prefecture contained 310 nanograms, followed by 300 nanograms in bonito caught in the central Sea of Japan, according to the team. The researchers said those concentration levels are comparable to levels found in fish living in contaminated waters like Tokyo Bay or off the Italian coast. In the fish caught in the South Pacific or around the Philippines, the team found only 24 to 50 nanograms of tin. (Source: *Environmental News Network*, April 8, 1997.)

TBT in Dolphins

The February 8, 1997 issue of *Science News* contained a report of dolphins found dead along the coast of Japan which had accumulated a variety of butyl tin compounds, the breakdown products of tributyl tin.

Over the previous decade, dolphins along the Atlantic and Gulf Coasts as well as in the Mediterranean Sea experienced several mass die-offs. In the tissues of the dead dolphins scientists found TBT. While the finding doesn't prove that tributyl tin killed the coastal-dwelling bottlenosed dolphins, there is other evidence that butyl tin compounds are potent immune system suppressors and they may have diminished the dolphins' ability to fight off the bacterial or viral infections thought to have caused the deaths. Exposure to TBT will continue because it persists in sediments and is still allowed on large vessels and aluminum hulls.

Before continuing the analysis on our web site, we will show you how to calculate the number of atoms or molecules of a substance in seawater using gold as an example. You can use this information later to convert PCB concentrations into the number of molecules per liter of seawater.

A Sample Calculation: Gold in Seawater

Gold is dissolved in seawater at a concentration of about 0.004 parts per billion (ppb). Intuitively, you probably know that 0.004 ppb is a very small amount. But how small? How much gold is in a liter of seawater?

To calculate the number of atoms of gold in seawater, you will need the following information:

- the concentration of gold (0.004 ppb)
- the atomic weight of gold (107.9 g mole^{-1})
- Avogadro's number (6.023 $\times$ 10^{23} atoms/mole)
- the *density* of full-strength (35 parts per thousand salinity) seawater (1.028 g ml^{-1} under "standard" conditions: 25°C and 1 atmosphere pressure).

Step 1: The first step in solving this problem is to state an equivalency:

By definition, 0.004 ppb is equivalent to 0.004 grams of gold per million kilograms (g 106 kg^{-1}) of seawater, or 0.004 micrograms per kilogram (mg·kg^{-1}).

Step 2: The second step is to find the number of moles of gold in a mass of water (we'll use kg for convenience). To do this, multiply the concentration of gold in seawater by the factor for converting grams to moles for gold. Note that 1 kg = 10^3 g.

$$0.004 \times 10^{-6}g/10^3g \times 1 \text{ mole}/107.9g$$
$$= 3.7 \times 10^{-11} \text{ mole}/10^3g$$

Step 3: We next need to determine the number of atoms of gold there are in 1 kg of seawater. Avogadro's number tells you how many atoms (ions, molecules, etc.) of an element or compound there are in one mole of that element. Thus, multiply the number you just calculated by the moles-to-atoms conversion factor.

$$3.7 \times 10^{-11} \text{ mole}/10^3g \times 6.023 \times 10^{23} \text{ atoms/mole}$$
$$= 2.2 \times 10^{13} \text{ atoms}/10^3g$$

Step 4: Thus, we now have determined the number of atoms of gold in a given mass of seawater. To calculate the number of atoms in a given volume of seawater, multiply the number of atoms of gold per 10^3 g of seawater by the density of seawater:

$$2.2 \times 10^{13} \text{ atoms}/10^3g \times 1.028 \text{ g/ml} = 2.3 \times 10^{10} \text{ atoms/ml}$$

Step 5: Use the ml-to-l conversion factor to convert your answer to atoms per liter.

$$2.3 \times 10^{10} \text{ atoms/ml} \times 1000 \text{ ml/l} = 2.3 \times 10^{13} \text{ atoms/l}$$

Now, go to our web site to continue this analysis.

Find the complete issue discussion and project at www.prenhall.com/oceanissues.

[1] Johnston, P. and I. McCrea. 1992. The effects of organochlorines on aquatic ecosystems. Greenpeace International, London.

[2] Lyke, M.L. Nov. 8, 1999. Toxin threatens a wonder of the northwest. *Washington Post*, p. A-9.

Bycatch: Dolphin-Safe Tuna[1] and Turtle-Safe Shrimp

- What is bycatch?
- What do consumers think "dolphin-safe" means?
- What is the exact meaning of "dolphin-safe"?
- What are the social, economic, and environmental costs and benefits of dolphin-safe fishing methods?
- What *is* an ecologically acceptable impact for commercial fishing?

Introduction

One of the main problems with large scale "harvesting" of wild marine organisms for human consumption is that most commercial fishing techniques are indiscriminate, that is, they cannot selectively capture only the target species. As a result, as much as 25% of the total global commercial catch is wasted or unused.

This quantity is known as "bycatch" and refers to undersized fish, low-value, and non-target species. It may include benthic organisms like sponges, worms, sea stars, crabs, etc., and also sharks, dolphins, whales, sea turtles, and even birds. Bycatch may die in nets or on longlines, or is often returned to the water dead or dying.

Shrimp Trawling

Among the most harmful fishing activities is trawling for shrimp. In addition to damaging the ocean bottom (trawling has been compared to clearcutting a forest!), as much as 90% of the trawl contents may be non-target and hence unused species, sometimes called "trash fish" by fishers, and at times including endangered sea turtles.

> Activity: Stop and consider how the phrase "trash fish" strikes you. Does it show respect for life?

Shrimp trawling is widespread and may cause extreme environmental damage. While as many as 25,000 boats ply U.S. waters and the U.S. imports wild-caught shrimp from nearly 40 countries, consumers are virtually unaware of the dimensions of its destructiveness. We'll discuss aspects of this issue on our web site.

In contrast to this ignorance of the environmental impact of shrimp trawling is a bycatch issue that most canned tuna consumers and others are well aware of—the capture of dolphins by tuna fishers.

In this issue we will focus on the multifaceted issue of bycatch in the tuna fishing industry and evaluate the costs and benefits of bycatch-reduction techniques.

Dolphins and Tuna in the ETP

The eastern tropical Pacific Ocean (ETP), an area of approximately 8 million square miles (21,000,000 km^2), is one of the world's richest sources of commercially important tunas. The ETP fishery for yellowfin tuna (*Thunnus albacares*), in fact, has been called one of the most important fisheries in the world[2]. Yellowfin and skipjack tunas (*Katsuwonus pelamis*) are mainstays of the canned light meat tuna industry. The ETP fishery for albacore (*Thunnus alalunga*), whose flesh is the basis of the white-meat tuna industry, is small by comparison.

Two methods have been widely used to catch yellowfin and skipjack tunas in large-scale fisheries in the ETP. In *school fishing* (Figure 1), a technique no longer practiced in the ETP, rugged commercial fishers used stout rods to catch tunas, which frequently bit unbaited hooks during their feeding frenzy. Worldwide, according to Bumblebee Seafoods, 40% of the world's commercial tuna are caught on pole and line.

A more productive method of catching yellowfin and skipjack tunas is *purse seining*. Globally, 30% of the

Figure 1

School fishing for tuna. (© Hulton-Deutsch Collection/CORBIS)

Figure 2

A tuna purse seiner with spread net. When the circle is complete, the bottom of the net is closed, trapping all sea creatures within.   *(Photos by Glenn Oliver/Visuals Unlimited)*

Figure 3

These photos of dead dolphins being hauled on board the Panamanian tuna boat "Maria Luisa" were captured from a video taken by a marine biologist who went undercover for five months on the boat to document the dolphin killings. The video from which these images were taken was broadcast on U.S. television and resulted in a huge public outcry.   *(AP LaserPhoto)*

world's commercial tuna are caught in purse seines. (Longlining, in which hooks are set at intervals along a horizontal line stretching for miles, accounts for 30% of world commercial tuna catch, essentially albacore, which are also caught commercially by trolling).

In purse seining (Figure 2), a school of fish is encircled by speed boats with a net that may be 2 km (1.2 mi.) long and 200 meters (660 ft) deep. A purse line attached to the bottom of the net is then pulled in, trapping the tunas and other organisms unfortunate enough to be in the same location. Vessels from 12 nations, including the U.S., purse seine in the ETP for tuna.

In the ETP, tunas frequently congregate around floating objects, such as tree trunks[3], and also along with two kinds of dolphins, northern offshore spotted (*Stenella attenuata*) and eastern spinner (*Stenella longirostris*), a fact discovered by the U.S. tuna fleet nearly three decades ago. This relationship is thought to benefit the tunas, which can easily follow dolphins and take advantage of the latter's superior prey-finding abilities. Setting nets around dolphins typically catches the largest tunas and is thus the more desirable method.

When modern tuna seiners enter an area, they can spot aggregations of tunas and dolphins fairly easily, especially by helicopter, because dolphins are noisy and disturb the sea surface. The netting process, which can take 2–3 hours, does not discriminate between the tunas and the dolphins, which stay together throughout the process. A number of dolphins can die due directly to entanglement and drowning (Figure 3), and more may die later due to the delayed effects of severe trauma. It is estimated that the purse-seine fishery for tuna killed more then 1.3 million eastern spotted dolphins in the ETP between 1959 and 1990. As many as 5 million dolphins were killed during the first 14 years of purse seining in the ETP.

Policies to Curb Dolphin Mortality

There have been several legislative and international attempts to curb the killing of dolphins during tuna seining. The first of these was an agreement reached with the Inter-American Tropical Tuna Commission (IATTC) in 1976 (but not funded until 1979). This program sought (1) to determine dolphin mortality, (2) to reduce it such that dolphin populations were not threatened and accidental killing was avoided, and (3) to maintain a high level of tuna production. The chief result of this effort was the placement of observers on one-third of all vessels fishing in the ETP. As a result, the first reliable estimates of dolphin mortality were made.

This was followed by a series of policies designed to lower dolphin mortality, including new provisions of the Marine Mammal Protection Act (MMPA) in 1988 and 1990; the Dolphin Protection Consumer Information Act of 1990; the Agreement for the Conservation of Dolphins of 1992 (more commonly called the La Jolla Agreement); the Panama Declaration of 1995; and the International Dolphin Conservation Program Act (IDCPA) of 1999.

These resulted in 100% observer coverage of ETP tuna seiners and established international limits of fewer than 5000 dolphins killed by 1999. Moreover, criteria were instituted for labeling canned tuna as "dolphin-safe". As we will see, the success of the "dolphin-safe" labeling program as a deterrent to killing dolphins is unsettled. However, as a marketing tool it is unequivocal: people buy the product. For 1996, domestic canned tuna sales approached $1 billion (Statistical Abstract of the U.S. 1998, p. 697, Table 1164)

There is no question that dolphin mortality has decreased in the ETP as a result of conservation measures. But the issue remains controversial and repercussions have been felt ecologically, economically, socially, and politically, as you will see below and on our web site.

The Issues

• What does "dolphin-safe" really mean?   The Dolphin Protection Consumer Information Act (DPCIA) of 1990 established minimum criteria for tuna labeled "dolphin safe" in the U.S. Essentially, for tuna caught from any vessel to be labeled "dolphin-safe" meant that intentional encirclement of dolphins did not occur.

A problem with this was that only about 20% of commercial tuna were caught in the ETP. There, enforcement of "dolphin-safe" capture techniques was fairly tight. However, the remaining purse-seined tuna catch was not subjected to the same stringent standards. In some cases, tuna were allowed to be designated "dolphin-safe" if the ship's skipper declared it so. Furthermore, the absence of safeguards meant that real dolphin-safe tuna, ersatz dolphin-safe tuna, and non-dolphin-safe tuna, could all be found on your grocery store shelf, labeled "dolphin-safe." This also placed U.S. and other ETP purse seiners under what many considered an undue burden and hindered their ability to compete fairly.

In April, 1999, then U.S. Commerce Secretary William Daley announced that "dolphin safe" could be used to designate any tuna harvested in the ETP if no dolphins were killed or seriously injured, even if encirclement of dolphins occurred.

This decision was denounced by a number of non-profit organizations, including the Sierra Club, Greenpeace, and the Humane Society of the U.S. Major U.S. tuna companies announced that they would continue to honor non-encirclement policies.

An alternative type of "dolphin safe" labeling has been devised by the non-profit organization Earthtrust. It awards the "Flipper Seal of Approval" to companies that meet a more stringent set of criteria (http://www.earth trust.org/fsareq.html)

• Have dolphin populations in the ETP been threatened by incidental capture in tuna purse seines? Are they now?   As we have seen, millions of dolphins have likely been killed in the ETP since the inception of dolphin encirclement. Today, that number has decreased significantly. To determine if a level of dolphin mortality threatens the stability of their populations, scientists must examine, among other data, the recruitment rate of the dolphin population. The **recruitment rate** is an estimate of the rate at which new individuals (i.e., recently born individuals) survive to enter the population. In this case, it provides policy makers and biologists with an estimate of the dolphin mortality that may be "acceptable," at least from the perspective of population stability. With respect to ETP dolphins, the question is whether mortalities caused by incidental catch in purse seines exceeds the recruitment rate.

We will examine this issue more fully on our web site.

• Do alternative methods reduce dolphin bycatch?   Two methods of purse-seining in the ETP are most common—*dolphin sets* (in which nets are dropped around schools of dolphins) and *log sets* (encircling floating objects such as trees, under which fish congregate). Dolphin sets typically kill 29 dolphins per 1,000 tons of tuna. Log sets kill fewer than one. Clearly, then, log sets reduce dolphin mortality, but they do so at the cost of much increased bycatch of other marine organisms. On our web site, we provide you with the data to analyze the magnitude and consequences of differences in bycatch between the two methods.

• What are the other issues?   Like most issues, the dolphin-tuna controversy has many dimensions. In Mexico, as many as 15,000 jobs in the tuna fishing and canning industry have been lost, and this loss has been attributed directly to the "dolphin-safe" issue.

Also, dolphins are relatively intelligent animals and are revered by many.

Now, proceed to our web site to continue analysis of this complex issue.

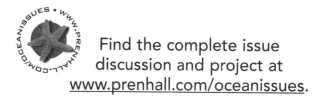

Find the complete issue discussion and project at www.prenhall.com/oceanissues.

[1] The "dolphin-safe" portion of this issue was written with Robert Young of Coastal Carolina University.

[2] Joseph, James. 1994. The tuna-dolphin controversy in the Eastern Pacific Ocean: biological, economic, and political impacts. *Ocean Devel. Inter. Law* 25: 1- 30.

[3] Surprisingly, enough such objects enter the ocean to be worthwhile to commercial fishers.

[4] Joseph, James. 1994. The tuna-dolphin controversy in the Eastern Pacific Ocean: biological, economic, and political impacts. *Ocean Devel. Inter. Law* 25: 1- 30.

[5] Summary Minutes of the 33rd Meeting of the Inter-American Tropical Tuna Commission, Managua, Nicaragua, October 11 – 14, 1976. IATTC, La Jolla CA.

NOTES:

NOTES:

NOTES:

NOTES: